ERIC R. KANDEL

THE DISORDERED MIND

Eric R. Kandel is a University Professor and the Fred Kavli Professor at Columbia University and a Senior Investigator at the Howard Hughes Medical Institute. The recipient of the 2000 Nobel Prize in Physiology or Medicine for his studies of learning and memory, he is the author of *In Search of Memory*, *The Age of Insight*, and *Reductionism in Art and Brain Science* and the coauthor of *Principles of Neural Science*, the standard textbook in the field.

ALSO BY ERIC R. KANDEL

Reductionism in Art and Brain Science: Bridging the Two Cultures

The Age of Insight: The Quest to Understand the Unconscious in Art, Mind, and Brain, from Vienna 1900 to the Present

In Search of Memory: The Emergence of a New Science of Mind

Principles of Neural Science (coauthor)

THE DISORDERED MIND

THE DISORDERED MIND

WHAT UNUSUAL BRAINS TELL US ABOUT OURSELVES

ERIC R. KANDEL

FARRAR, STRAUS AND GIROUX   NEW YORK

Farrar, Straus and Giroux
120 Broadway, New York 10271

Printed in the United States of America
Published in 2018 by Farrar, Straus and Giroux
First paperback edition, 2019

The Library of Congress has cataloged the hardcover edition as follows:
Names: Kandel, Eric R., author.
Title: The disordered mind : what unusual brains tell us about ourselves / Eric R. Kandel.
Description: First edition. | New York : Farrar, Straus and Giroux, 2018. | Includes
bibliographical references and index.
Identifiers: LCCN 2017049274 | ISBN 9780374287863 (hardcover)
Subjects: | MESH: Mental Disorders—physiopathology | Neuropsychiatry—methods |
Mental Processes—physiology | Psychophysiologic Disorders—physiopathology
Classification: LCC RC454 | NLM WM 102 | DDC 616.89—dc23
LC record available at https://lccn.loc.gov/2017049274

Paperback ISBN: 978-0-374-53844-6

3 5 7 9 10 8 6 4 2

For Denise,

My constant companion, my strongest critic,

and my continuous source of inspiration

The mind is like an iceberg, it floats with one-seventh of its bulk above water. —Sigmund Freud

CONTENTS

THE DISORDERED MIND

INTRODUCTION

I have spent my entire career trying to understand the inner workings of the brain and the motivation for human behavior. Having escaped from Vienna as a young boy soon after Hitler occupied it, I was preoccupied with one of the great mysteries of human existence: How can one of the most advanced and cultured societies on earth turn its efforts so rapidly toward evil? How do individuals, when faced with a moral dilemma, make choices? Can the splintered self be healed through skilled human interaction? I became a psychiatrist in hopes of understanding and acting on these difficult problems.

As I began to appreciate the elusiveness of the problems of the mind, however, I turned to questions that could be answered more definitively through scientific research. I focused on small collections of neurons in a very simple animal, and eventually discovered some of the fundamental processes underlying elementary forms of learning and memory. While I have enjoyed my work a great deal and it has been amply appreciated by others, I realize that my findings represent but a small advance in the quest to understand the most complex entity in the universe—the human mind.

This pursuit has animated philosophers, poets, and physicians since the dawn of humankind. Engraved on the entrance of the Temple of Apollo at Delphi was the maxim "Know thyself." Ever since Socrates and Plato first reflected on the nature of the human mind, serious thinkers of every generation have sought to understand the thoughts, feelings, behavior, memories, and creative powers that make us who we are. For earlier generations, this quest was restricted to the intellectual framework of philosophy, as embodied in the seventeenth-century French philosopher René Descartes's pronouncement "I think, therefore I am." Descartes's guiding idea was that our mind is separate from, and functions independently of, our body.[1]

One of the great steps forward in the modern era was the realization that Descartes had it backward: in actuality, "I am, therefore I think." This reversal came about in the late twentieth century, when a school of philosophy that was concerned with mind, a school led importantly by people such as John Searle and Patricia Churchland, merged with cognitive psychology,[2] the science of mind, and both then merged with neuroscience, the science of the brain. The result was a new, biological approach to mind. This unprecedented scientific study of mind is based on the principle that our mind is a set of processes carried out by the brain, an astonishingly complex computational device that constructs our perception of the external world, generates our inner experience, and controls our actions.

The new biology of mind is the last step in the intellectual progression that began in 1859 with Darwin's insights into the evolution of our bodily form. In his classic book *On the Origin of Species*, Darwin introduced the idea that we are not unique beings created by an all-powerful God, but instead are biological creatures that have evolved from simpler animal ancestors and share with them a combination of instinctual and learned behavior. Darwin elaborated on this idea in his 1872 book, *The Expression of the Emotions in Man and Animals*,[3] in which he presented an even more radical and profound idea: that our mental processes evolved from animal ancestors in much the same way that our morphological features did. That is, our mind is not ethereal; it can be explained in physical terms.

Brain scientists, myself included, soon realized that if simpler animals exhibit emotions similar to ours, such as fear and anxiety in response to threats of bodily harm or diminished social position, we should be able to study aspects of our own emotional states in those animals. It subsequently became clear from studies of animal models that, much as Darwin had predicted, even our cognitive processes, including primitive forms of consciousness, evolved from our animal ancestors.

The fact that we share aspects of our mental processes with simpler animals and can therefore study the workings of the mind on an elementary level in those animals is fortunate, because the human brain is astonishingly complex. That complexity is most evident—and most mysterious—in our awareness of self.

Self-awareness leads us to question who we are and why we exist. Our myriad creation mythologies—the stories each society tells about its origins—arose from this need to account for the universe and our place in it. Seeking answers to these existential questions is an important part of what defines us as human beings. And seeking answers to how the intricate interactions of brain cells give rise to consciousness, to our awareness of self, is the great remaining mystery in brain science.

How does human nature arise from the physical matter of the brain? The brain can achieve consciousness of self and can perform its remarkably swift and accurate computational feats because its 86 billion nerve cells—its neurons—communicate with one another through very precise connections. During the course of my career, my colleagues and I have been able to show in a simple invertebrate marine animal, *Aplysia*, that these connections, known as synapses, can be altered by experience. This is what enables us to learn, to adapt to changes in our surroundings. But the connections among neurons can also be altered by injury or disease; moreover, some connections may fail to form normally during development, or even to form at all. Such cases result in disorders of the brain.

Today, as never before, the study of brain disorders is giving us new insights into how our mind normally functions. What we are learning about autism, schizophrenia, depression, and Alzheimer's disease, for example, can help us understand the neural circuits involved in social interactions, in thoughts, feelings, behavior, memory, and creativity just as surely as studies of those neural circuits can help us understand brain disorders. In a larger sense, much as the components of a computer reveal their true functions when they break down, so the functions of the brain's neural circuits become dramatically clear when they falter or fail to form correctly.

This book explores how the processes of the brain that give rise to our mind can become disordered, resulting in devastating diseases that haunt humankind: autism, depression, bipolar disorder, schizophrenia, Alzheimer's disease, Parkinson's disease, and post-traumatic stress disorder. It explains how learning about these disordered processes is essential for improving our understanding of the normal workings of the brain, as well as for finding new treatments for the disorders. It also illustrates that we can enrich our understanding of how the brain works by examining normal variations in brain function, such as how the brain differentiates during development to determine our sex and our gender identity. Finally, the book shows how the biological approach to mind is beginning to unravel the mysteries of creativity and of consciousness. We see, in particular, remarkable instances of creativity in people with schizophrenia and bipolar disorder and find that their creativity arises from the same connections between brain, mind, and behavior present in everyone. Modern studies of consciousness and its disorders suggest that consciousness is not a single, uniform function of the brain; instead, it is different states of mind in different contexts. Moreover, as earlier scientists discovered and as Sigmund Freud had emphasized, our conscious perceptions, thoughts, and actions are informed by unconscious mental processes.

In a larger sense, the biological study of mind is more than a scientific inquiry holding great promise for expanding our understanding of the brain and devising new therapies for people who have disorders of the brain. Advances in the biology of mind offer the possibility of a new humanism, one that merges the sciences, which are concerned with the natural world, and the humanities, which are concerned with the meaning of human experience. This new scientific humanism, based in good part on biological insights into differences in brain function, will change fundamentally the way we view ourselves and one another. Each of us already *feels* unique, thanks to our consciousness of self, but we will actually have biological confirmation of our individuality. This, in turn, will lead to new insights into human nature and to a deeper understanding and appreciation of both our shared and our individual humanity.

1

WHAT BRAIN DISORDERS CAN TELL US ABOUT OURSELVES

The greatest challenge in all of science is to understand how the mysteries of human nature—as reflected in our individual experience of the world—arise from the physical matter of the brain. How do coded signals, sent out by billions of nerve cells in our brain, give rise to consciousness, love, language, and art? How does a fantastically complex web of connections give rise to our sense of identity, to a self that develops as we mature yet stays remarkably constant through our life experiences? These mysteries of the self have preoccupied philosophers for generations.

One approach to solving these mysteries is to reframe the question: What happens to our sense of self when the brain does not function properly, when it is beset by trauma or disease? The resulting fragmentation or loss of our sense of self has been described by physicians and lamented by poets. More recently, neuroscientists have studied how the self comes undone when the brain is under assault. A famous example is Phineas Gage, the nineteenth-century railway worker whose personality changed dramatically after an iron rod pierced the front of his brain. Those who had known him before his injury said simply, "Gage is no longer Gage."

This approach implies a "normal" set of behaviors, both for an individual and for people in general. The dividing line separating "normal" and "abnormal" has been drawn in different places by different societies throughout history. People with mental differences have sometimes been seen as "gifted" or "holy," but more frequently they have been treated as "deviant" or "possessed" and subjected to terrible cruelty and stigmatization. Modern psychiatry has attempted to describe and catalogue mental disorders, but the migration of various behaviors across the line separating the normal from the disordered is a testament to the fact that the boundary is indistinct and mutable.

All of these variations in behavior, from those considered normal to those considered abnormal, arise from individual variations in our brains. In fact, every activity we engage in, every feeling and thought that gives us our sense of individuality, emanates from our brain. When you taste a peach, make a difficult decision, feel melancholy, or experience a rush of joyous emotion when looking at a painting, you are relying entirely on the brain's biological machinery. Your brain makes you who you are.

You're probably confident that you experience the world as it is—that the peach you see, smell, and taste is exactly as you perceive it. You rely on your senses to give you accurate information so that your perceptions and actions are grounded in an objective reality. But that's only partly true. Your senses do provide the information you need to act, but they don't present your brain with an objective reality. Instead, they give your brain the information it needs to *construct* reality.

Each of our sensations emerges from a different system of the brain, and each system is fine-tuned to detect and interpret a particular aspect of the external world. Information from each of the senses is gathered by cells designed to pick up the faintest sound, the slightest touch or movement, and this information is carried along a dedicated pathway to a region of the brain that specializes in that particular sense. The brain then analyzes the sensations, engaging relevant emotions and memories of past experience to construct an internal representation of the outside world. This self-generated reality—in part unconscious, in part conscious—guides our thoughts and our behavior.

Ordinarily, our internal representation of the world overlaps to a great degree with everyone else's, because our neighbor's brain has evolved to work in the same way as our own; that is, the same neural circuits underlie the same mental processes in every person's brain. Take language, for example: the neural circuits responsible for expression of language are located in one area of the brain, while the circuits responsible for comprehension of language are located in another area. If during development those neural circuits fail to form normally, or if they are disrupted, our mental processes for language become disordered and we begin to experience the world differently from other people—and to act differently.

Disruptions of brain function can be both frightening and tragic, as anyone who has witnessed a grand mal seizure or seen the anguish of a deep depression can tell you. The effects of extreme mental illness can be devastating to individuals and their families, and the global suffering from these diseases is immeasurable. But some disruptions of typical brain circuitry can confer benefits and affirm a person's individuality. In fact, a surprising number of people who suffer from what one might see as a disorder would choose not to eradicate that aspect of themselves. Our sense of self can be so powerful and essential that we are reluctant to relinquish even those portions of it that cause us to suffer. Treatment of these conditions too often compromises the sense of self. Medications can deaden our will, our alertness, and our thought processes.

Brain disorders provide a window into the typical healthy brain. The more scientists and clinicians learn about brain disorders—from observing patients and from neuroscientific and genetic research—the more they understand about how the mind works when all brain circuits are functioning robustly, and the more likely they are to be able to develop effective treatments when some of those circuits fail. The more we learn about unusual minds, the more likely we are as individuals and as a society to understand and empathize with people who think differently and the less likely we are to stigmatize or reject them.

PIONEERS IN NEUROLOGY AND PSYCHIATRY

Until about 1800, only disorders that resulted from visible damage to the brain, as seen at autopsy, were considered medical disorders; these disorders were labeled neurological. Disorders of thought, feelings, and mood, as well as drug addiction, did not appear to be associated with detectable brain damage and, as a result, were considered to be defects in a person's moral character. Treatments for these "weak-minded" people were designed to "toughen them up" by isolating them in asylums, chaining them to the walls, and exposing them to deprivations or even torture. Not surprisingly, this approach was medically fruitless and psychologically destructive.

In 1790 the French physician Philippe Pinel formally founded the field we now call psychiatry. Pinel insisted that psychiatric disorders are not moral disturbances but medical diseases, and that psychiatry should be considered a subdiscipline of medicine. At Salpêtrière, Paris's large psychiatric hospital, Pinel freed the mental patients from their chains and introduced humane, psychology-oriented principles that were a forerunner of present-day psychotherapy.

Pinel argued that psychiatric disorders strike people who have a hereditary predisposition and who are exposed to excessive social or psychological stress. This view is remarkably close to the view of mental illness that we hold today.

Although Pinel's ideas had a great moral impact on the field of psychiatry by humanizing the treatment of patients, no further progress was made in understanding psychiatric disorders until the early twentieth century, when the great German psychiatrist Emil Kraepelin founded modern scientific psychiatry. Kraepelin's influence cannot be overstated, and I will weave his story through this book as it weaves through the history of neurology and psychiatry.

Kraepelin was a contemporary of Sigmund Freud, but whereas Freud believed that mental illnesses, although based in the brain, are acquired through experience—often a traumatic experience in early childhood—Kraepelin held a very different view. He thought that all mental illnesses have a biological origin, a genetic basis. As a result, he reasoned, psychiatric illnesses could be distinguished from one another much as other medical illnesses are: by observing their initial manifestations, their clinical courses over time, and their long-term outcomes. This belief led Kraepelin to establish a modern system for classifying mental illness, a system still in use today.

Kraepelin was inspired to take a biological view of mental illnesses by Pierre Paul Broca and Carl Wernicke, two physicians who first illustrated that we can gain remarkable insights into ourselves by studying disorders of the brain. Broca and Wernicke discovered that specific neurological disorders can be traced to specific regions of our brain. Their advances led to the realization that the mental functions underlying normal behavior can also be localized to specific regions and sets of regions of the brain, thus laying the groundwork for modern brain science.

In the early 1860s Broca noticed that one of his patients, a man named Leborgne, who suffered from syphilis, had a peculiar language deficit. Leborgne could understand language perfectly well, but he couldn't make himself understood. He could take in what someone told him, as evidenced by his ability to follow instructions to the letter, but when he tried to speak, only unintelligible mumbles came out. The man's vocal cords weren't paralyzed—he could easily hum a tune—but he could not express himself in words. Nor could he express himself through writing.

After Leborgne died, Broca examined his brain, looking for clues to his affliction. He found a region in the forward part of the left hemisphere that appeared blighted by disease or injury. Broca eventually encountered eight additional patients with the same difficulty producing language and found that they all had damage in the same area on the left side of the brain, a region that became known as Broca's area (fig. 1.1). These findings led him to conclude that our ability to speak resides in the left hemisphere of the brain, or as he put it, "We speak with the left hemisphere."[1]

In 1875 Wernicke observed the mirror image of Leborgne's defect. He encountered a patient whose words flowed freely but who could not understand language. If Wernicke told him to "Put object A on top of object B," the man would have no idea what he was being asked to do. Wernicke tracked this deficit in language comprehension to damage in the back of the left hemisphere, a region that became known as Wernicke's area (fig. 1.1).

Wernicke had the great insight to realize that complex mental functions like language do not reside in a single region of the brain but instead involve multiple, interconnected brain regions. These circuits form the neural "wiring" of our brain. Wernicke demonstrated not only that comprehension and expression are processed separately but that they are connected to each other by a pathway known as the *arcuate fasciculus*. The information we obtain from reading is transmitted from our eyes to the visual cortex, and the information from hearing is sent from our ears to the auditory cortex. Information from these two cortical areas then converges in Wernicke's area, which translates it into a neural code for understanding language. Only then does the information proceed to Broca's area, enabling us to express ourselves (fig. 1.1).

Wernicke predicted that someday, someone would find a disorder of language that involves simply a disconnect between the two areas. This proved to be the case: people with damage to the arcuate pathway connecting the two areas can understand language and express language, but the two functions operate independently. This is a bit like a presidential press conference: information comes in, information goes out, but there is no logical connection between them.

Scientists now think that other complex cognitive skills also require the participation of several quite distinct but interconnected regions of the brain.

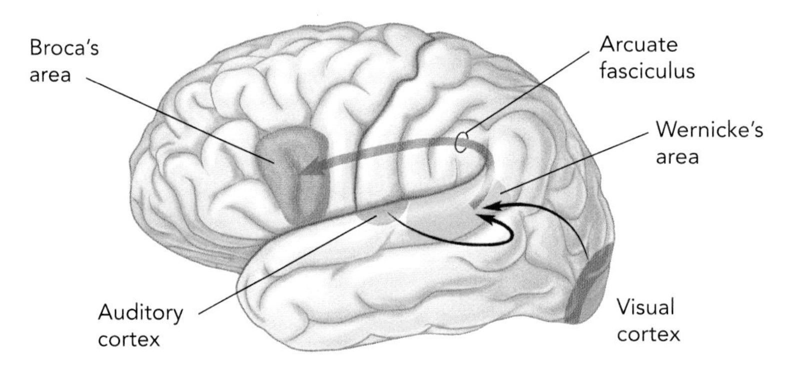

Figure 1.1. The anatomical pathway for language comprehension (Wernicke's area) and expression (Broca's area). The two areas are connected by the arcuate fasciculus.

Although the circuitry for language has proved to be even more complex than Broca and Wernicke realized, their initial discoveries formed the basis of our modern view of the neurology of language and, by extension, our view of neurological disorders. Their emphasis on location, location, location resulted in major advances in the diagnosis and treatment of neurological disease. Moreover, the damage typically caused by neurological diseases is easily visible in the brain, making them far easier to identify than most psychiatric disorders, in which the damage is much subtler.

The search for localization of function in the brain was enhanced dramatically in the 1930s and '40s by Canada's renowned neurosurgeon Wilder Penfield, who operated on people suffering from epilepsy caused by scar tissue that had formed in the brain after a head injury. Penfield was seeking to elicit an aura, the sensation many epileptic patients experience before a seizure. If successful, he would have a good idea of which tiny bit of the brain to remove in order to relieve his patients' seizures without damaging other functions, such as language or the ability to move.

Penfield's patients were awake during the operation—the brain has no pain receptors—so they could tell him what they were experiencing when he stimulated various areas in their brain. Over the next several years, in the course of nearly four hundred operations, Penfield mapped the regions of our brain that are responsible for the sensations of touch, vision, and hearing and for the movements of specific parts of our body. His maps of sensory and motor function are still used today.

What was truly amazing was Penfield's discovery that when he stimulated the temporal lobe, the part of the brain that is just above the ear, his patient might suddenly say, "Something is coming back to me as if it is a memory. I hear sounds, songs, parts of symphonies." Or, "I hear the lullaby my mother used to sing to me." Penfield began to wonder if it were possible to locate a mental process as complex and mysterious as memory to specific regions in the physical brain. Eventually, he and others determined that it is.

NEURONS: THE BUILDING BLOCKS OF THE BRAIN

Broca's and Wernicke's discoveries revealed *where* in the brain certain mental functions are located, but they stopped short of explaining *how* the brain carries them out. They were unable to answer basic questions such as, What is the biological makeup of the brain? How does it function?

Biologists had already established that the body is composed of discrete cells, but the brain appeared to be different. When scientists looked through their microscopes at brain tissue, they saw a tangled mess that seemed to have no beginning and no end. For this reason, many scientists thought the nervous system was a single, continuous web of interconnected tissue. They weren't sure there was such a thing as a discrete nerve cell.

Then, in 1873, an Italian physician named Camillo Golgi made a discovery that would revolutionize scientists' understanding of the brain. He injected silver nitrate or potassium dichromate into brain tissue and observed that, for reasons we still don't understand, a tiny fraction of the cells took up the stain and turned a distinctive black color. Out of an impenetrable block of neural tissue, the fine and elegant structure of individual neurons was suddenly thrown into high relief (fig. 1.2).

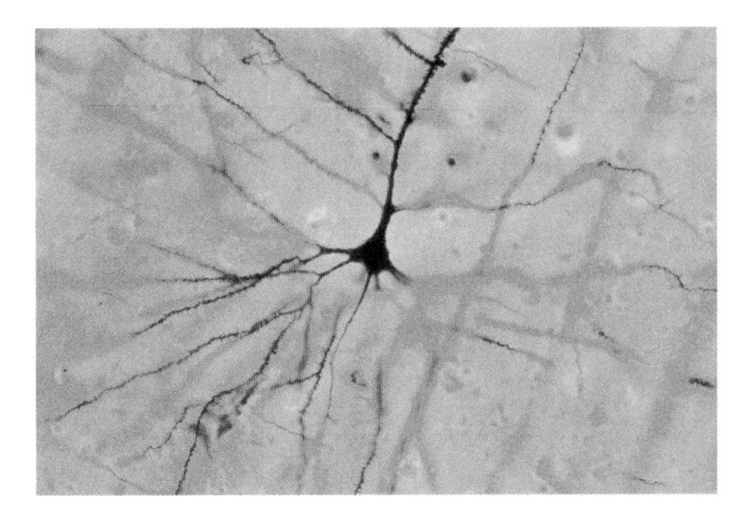

Figure 1.2. Golgi stain

The first scientist to take advantage of Golgi's discovery was a young Spaniard named Santiago Ramón y Cajal. In the late 1800s Cajal applied Golgi's stain to brain tissue from newborn animals. This was a wise move: early in development the brain has fewer neurons, and their shape is simpler, so they are easier to see and examine than neurons in a mature brain. Using Golgi's stain in the immature brain, Cajal could identify isolated cells and study them one at a time.

Cajal saw cells that resembled the sprawling canopies of ancient trees, others that ended in compact tufts, and still others that sent branches arcing into unseen regions of the brain—shapes that were completely different from the simple, well-defined shapes of other cells in the body. In spite of this startling diversity, Cajal determined that each neuron has the same four principal anatomical components (fig. 1.3): the cell body, the dendrites, the axon, and the presynaptic terminals, which end in what are now known as synapses. The main component of the neuron is the cell body, which contains the nucleus (the repository of the cell's genes) and the majority of the cytoplasm. The multiple, thin extensions from the cell body, which look like the slender branches of a tree, are the dendrites. Dendrites receive information from other nerve cells. The single thick extension from the cell body is the axon, which can be several feet long. The axon transmits information to other cells. At the end of the axon are the presynaptic terminals. These specialized structures form synapses with the dendrites of target cells and transmit information to them across a small gap known as the synaptic cleft. Target cells may be neighboring cells, cells in another region of the brain, or muscle cells at the periphery of the body.

Eventually, Cajal united these four principles in a theory now called

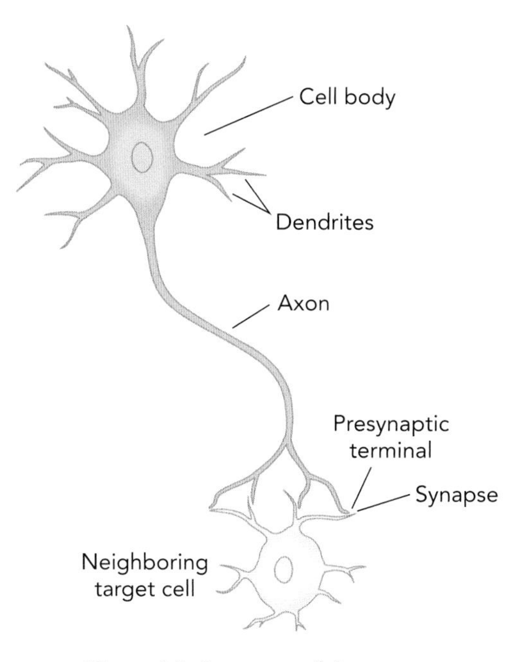

Figure 1.3. Structure of the neuron

the Neuron Doctrine (fig. 1.4). The first principle is that each neuron is a discrete element that serves as the fundamental building block and signaling unit of the brain. The second is that neurons interact with one another only at the synapses. In this way, neurons form the intricate networks, or neural circuits, that enable them to communicate information from one cell to another. The third principle is that neurons form connections only with particular target neurons at particular sites. This *connection specificity* accounts for the astonishingly precise circuitry that underlies the complex tasks of perception, action, and thought. The fourth principle, which derives from the first three, is that information flows in one direction only—from the den-

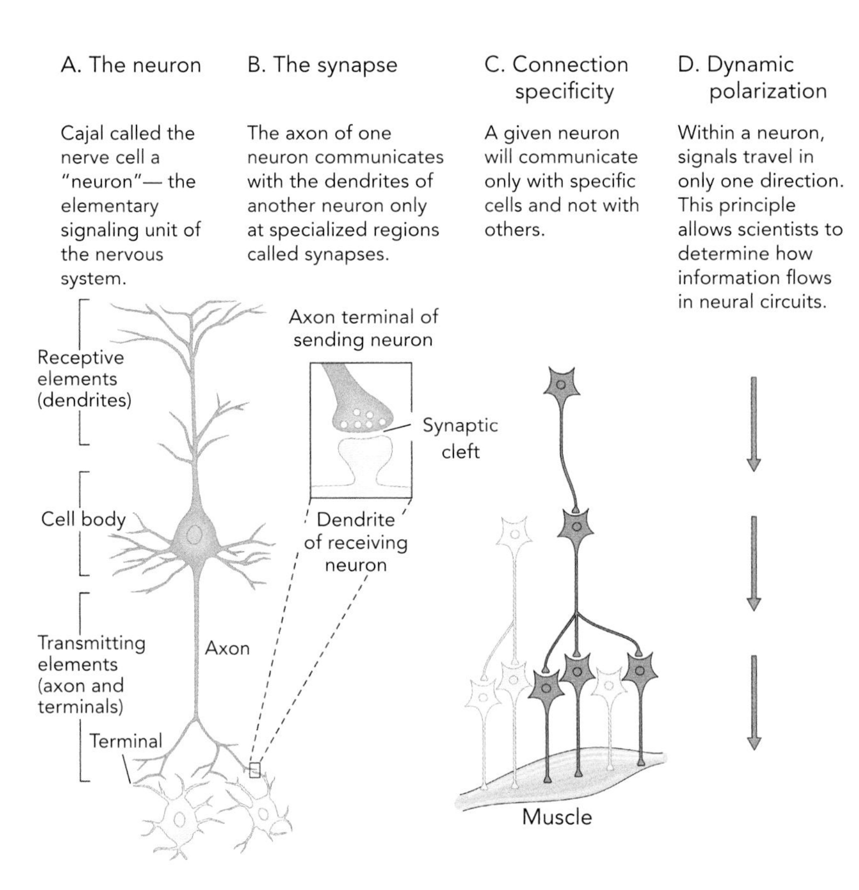

Figure 1.4. The four principles of Cajal's Neuron Doctrine

drites to the cell body to the axon, then along the axon to the synapse. We now call this flow of information in the brain the principle of *dynamic polarization.*

Cajal's ability to look through a microscope at a fixed array of neurons and imagine how the nervous system works was an extraordinary feat of scientific intuition. In 1906 he and Golgi were awarded the Nobel Prize in Physiology or Medicine—Golgi for his stain and Cajal for using it to establish the structure and function of neurons. Amazingly, Cajal's insights have held firm from 1900 to the present.

THE SECRET LANGUAGE OF NEURONS

For neurons to process information, and thus to instruct behavior, they need to communicate with other neurons and with the rest of the body. This is an absolute necessity for the brain to function properly. But how do neurons speak to one another? It wasn't until years later that the answer to this question began to emerge.

In 1928 Edgar Adrian, a pioneer in the electrophysiological study of the nervous system and recipient of the 1932 Nobel Prize in Physiology or Medicine, surgically exposed one of the many small nerves, or bundles of axons, in the neck of an anesthetized rabbit. He then removed all but two or three of the axons and placed an electrode on the remaining ones. Adrian observed a flurry of electrical activity every time the rabbit took a breath. He attached a loudspeaker to the electrode and immediately began to hear clicking noises, a fast rapping similar to Morse code. The clicking noise was an electrical signal, an *action potential,* the fundamental unit of neural communication. Adrian was listening in on the language of neurons.

What produced the action potentials that Adrian heard? The inside of the membrane that surrounds a neuron and its axon has a slightly negative electric charge relative to the outside. This charge results from an unequal distribution of ions—electrically charged atoms—on either side of the cell membrane. Because of this unequal distribution of ions, each neuron is like a tiny battery, storing electricity that can be released at any moment.

When something excites a neuron—whether it's a photon of light, a sound wave, or the activity of other neurons—microscopic gates called *ion channels* open up all over its surface, allowing the charged ions to rush across the membrane in both directions. This free flow of ions reverses the electrical polarity of the cell membrane, switching the charge inside the neuron from negative to positive and releasing the neuron's electrical energy.

The rapid discharge of energy causes the neuron to generate an action potential. This electrical signal propagates rapidly along the neuron, from its cell body to the tip of its axon. When scientists say neurons in a particular region of the brain are active, they mean that the neurons are firing action potentials. Everything we see, touch, hear, and think begins with these little spikes of electricity racing from one end of a neuron to the other.

Adrian next recorded electrical signals from individual axons in the optic nerve of a toad. He amplified the signals so that they could be displayed on an early version of the oscilloscope as a two-dimensional graph. In this way, he discovered that the action potentials in any given neuron maintain a fairly consistent size, shape, and duration. They are always the same little spike of voltage. He also found that a neuron's response to a stimulus is all or none: the neuron either generates a full-blown action potential or it doesn't fire one at all. Once initiated, an action potential travels without fail from the dendrites of the receiving cell to the cell body and along its axon to the synapse. This is quite a feat in, say, a giraffe, which has axons that begin in the spinal cord and extend several meters to the muscles at the end of its leg.

The fact that action potentials are all-or-none events raises two interesting questions. First, how does a neuron that responds to sensory stimuli report differences in the intensity of a stimulus? How does it distinguish a light touch from a heavy blow, or a dim light from a bright one? And second, do neurons carrying information from different senses—sight, touch, taste, hearing, or smell—use different types of signals?

Adrian found that a neuron signals intensity not by changing the strength or duration of its action potentials but by varying the frequency with which it fires them. A weak stimulus causes the cell to fire only a few action potentials, while a strong stimulus produces

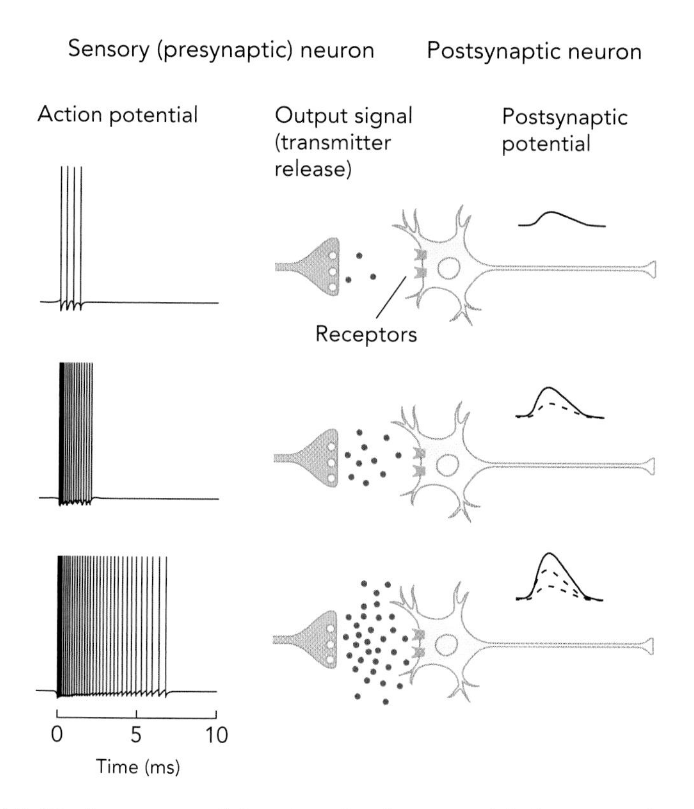

Figure 1.5. The frequency and duration of action potentials determine the strength of the neuron's downstream chemical signal.

much more frequent firing. Moreover, he could gauge the duration of the stimulus by monitoring the duration of the firing of action potentials (fig. 1.5).

Adrian went on to record action potentials from neurons in the eyes, skin, tongue, and ears to see whether they were different. He found that the signals are similar, irrespective of where they come from or what kind of sensory information they convey. What distinguishes sight from touch and taste from hearing is the particular neuronal pathway carrying the signal and its destination. Each type of sensory information is carried along its own neural pathway to a corresponding region in the brain.

How does an action potential in one neuron spark an action potential in the next neuron in the circuit? Two young British scientists, Henry

Dale and William Feldberg, observed that when an action potential reaches the end of the axon in the sending, or presynaptic, cell, something astonishing happens: the cell releases a spurt of chemicals into the synaptic cleft. These chemicals, now known as *neurotransmitters*, cross the synaptic cleft and bind to receptors on the dendrites of the target, or postsynaptic, cell. Each neuron sends information by making thousands of synaptic connections with its target cells and in turn receives information through thousands of connections from other neurons. The receiving neuron then adds up all the signals it has received through these connections, and if they are strong enough, the neuron translates them into a new action potential, a new all-or-none electrical signal that is transmitted to all the target cells that the receiving neuron makes contact with. The process is then repeated. In this way neurons can relay information almost instantaneously to other neurons and to muscle cells, even over long distances.

Alone, this simple computation might not seem very impressive, but when hundreds or thousands of neurons form circuits carrying signals from one part of the brain to another, the end result is perception, movement, thought, and emotion. The computational nature of the brain provides us with both a road map and a logic for analyzing disorders of the brain. That is, by analyzing glitches in neural circuits, we can begin to probe the mysteries of the brain—to figure out how electrical circuits generate perception, memory, and consciousness. As a corollary, brain disorders give us a way to see how the processes of the brain create mind and how most of our other experiences and behavior are rooted in this computational marvel.

THE DIVIDE BETWEEN PSYCHIATRY AND NEUROLOGY

Despite the many advances in brain science in the nineteenth century—advances that formed the foundation of modern neurology—psychiatrists and addiction researchers did not focus on the anatomy of the brain. Why didn't they?

For a very long time psychiatric and addictive disorders were viewed as fundamentally different from neurological disorders. When pathologists examined the brain of a patient at autopsy and found obvious damage, as they did in cases of stroke, head trauma, or syphilis and other infections of the brain, they classified the disorder as biological, or neurological. When they failed to detect clearly visible anatomical damage, they classified the disorder as functional, or psychiatric.

Pathologists were struck by the fact that most psychiatric disorders—namely, schizophrenia, depression, bipolar disorder, and anxiety states—did not produce visibly dead cells or holes in the brain. Since they did not see any obvious damage, they assumed that these disorders were either extracorporeal (disorders of mind rather than the body) or too subtle to detect.

Because psychiatric and addictive disorders did not produce obvious damage in the brain, they were considered to be behavioral in nature and thus essentially under the individual's control—the moralistic, nonmedical view that Pinel deplored. This view led psychiatrists to conclude that the social and functional determinants of mental disorders act on a different "level of the mind" than do the biological determinants of neurological disorders. The same was held to be true, at that time, of any deviation from the accepted norms of heterosexual attraction, feeling, and behavior.

Many psychiatrists considered the brain and mind to be separate entities, so psychiatrists and addiction researchers did not look for a connection between their patients' emotional and behavioral difficulties and the dysfunction or variation of neural circuits in the brain. Thus, for decades psychiatrists had difficulty seeing how the study of electrical circuits could help them explain the complexity of human behavior and consciousness. In fact, it was customary as late as 1990 to classify psychiatric illnesses as either organic or functional, and some people still use this outdated terminology. Descartes's mind-body dualism has proved hard to shake because it reflects the way we experience ourselves.

MODERN APPROACHES TO BRAIN DISORDERS

The new biology of mind that emerged in the late twentieth century is based on the assumption that all of our mental processes are mediated by the brain, from the unconscious processes that guide our movements as we hit a golf ball to the complex creative processes that underlie the composition of a piano concerto to social processes that allow us to interact with one another. As a result, psychiatrists now see our mind as a series of functions carried out by the brain, and they view all mental disorders, both psychiatric and addictive, as brain disorders.

This modern view derives from three scientific advances. The first was the emergence of a genetics of psychiatric and addictive disorders pioneered by Franz Kallmann, a German-born psychiatrist who immigrated to the United States in 1936 and worked at Columbia University. Kallmann documented the role of heredity in psychiatric disorders such as schizophrenia and bipolar disorder, thereby showing that they are indeed biological in nature.

The second advance was brain imaging, which has begun to show that the various psychiatric disorders involve distinct systems in the brain. It is now possible, for example, to detect some of the areas of the brain that function abnormally in people with depression. In addition, imaging has allowed researchers to watch the action of drugs on the brain and even to see the changes that result from treating patients with drugs or with psychotherapy.

The third advance was the development of animal models of disease. Scientists create animal models by manipulating the animals' genes and then observing the effects. Animal models have proven invaluable in studies of psychiatric disorders, providing insights into how genes, the environment, and the interaction of the two can disrupt brain development, learning, and behavior. Animal models, such as mice, are particularly useful for studying learned fear or anxiety because these states occur naturally in animals. But mice can also be used to study depression or schizophrenia by inserting into their brain altered genes that have been shown to contribute to depression or schizophrenia in people.

Let us first consider the genetics of mental disorders, then the imaging of brain functions, and finally animal models.

GENETICS

For all its wonders, the brain is an organ of the body—and like all biological structures, it is built and regulated by genes. Genes are distinct stretches of DNA that have two remarkable qualities: they provide cells with instructions for how to start an organism anew, and they are handed down from one generation to the next, thereby transferring those instructions to the organism's offspring. Each of our genes provides a copy of itself to almost every cell in our body, as well as to generations that succeed us.

We all have about twenty-one thousand genes, and roughly half of them are expressed in the brain. When we say a gene is "expressed," we mean that it is turned on, that it is busy directing the synthesis of proteins. Each gene encodes—that is, issues the instructions for making—a particular protein. Proteins determine the structure, function, and other biological characteristics of every cell in our body.

Genes generally replicate reliably, but when one doesn't, a mutation results. This alteration in a gene can occasionally prove beneficial to an organism, but it can also result in the overproduction, loss, or malfunction of the protein that the particular gene encodes, thus compromising cell structure and function and possibly leading to disorders.

Each of us has two copies of each gene, one from our mother and one from our father. The pairs of genes are arranged in precise order along twenty-three pairs of chromosomes. As a result, scientists can identify each gene by its location, or locus, on a specific chromosome.

The maternal and paternal copies of each gene are referred to as *alleles*. The two alleles of a particular gene usually differ slightly: that is, each one consists of a particular sequence of *nucleotides*, the four molecules that make up the code of DNA. Thus, the sequence of nucleotides in the genes you inherit from your mother is not exactly the same as the sequence of nucleotides in the genes you inherit from your father. Moreover, the nucleotide sequences you inherit are not exact copies of your parents' sequences; they contain some differences that occurred by chance when the gene was copied from your parent to you. These differences lead to variations in appearance and behavior.

In spite of the many variations that give us our sense of individuality, the genetic makeup, or *genome*, of any two people is more than 99 percent identical. The difference between them results from these chance variations in one or more of the genes they inherited from their parents (although there are rare exceptions, which we will touch on in chapter 2).

If almost every cell in our body contains the instructions for every other cell, then how is it that one cell becomes a kidney cell, while another becomes part of the heart? Or, in the brain, how does one cell become a hippocampal neuron, involved in memory, and another a spinal motor neuron, involved in the control of movement? In each instance, a distinct set of genes in the progenitor cell was activated, setting in motion the machinery that gave that cell its particular identity. Which particular set of genes is activated depends upon the interaction of molecules inside the cell and the interaction of the cell with both its neighboring cells and with the organism's external environment. We have a finite number of genes, but the turning on and off of different genes at different times gives rise to an almost infinite complexity.

To fully understand a brain disorder, scientists try to pinpoint the underlying genes and then understand how variations in those genes, interacting with the environment, bring about the disorder. With a basic knowledge of what has gone wrong, we can begin to figure out ways of intervening to prevent or ameliorate the disorder.

Genetic studies of families, beginning with those done by Kallmann in the 1940s, show just how pervasive genetic influences are in psychiatric disorders (table 1). We refer to genetic "influences" because the inheritance of psychiatric disorders is complex: there is no single gene that causes schizophrenia or bipolar disorder. What Kallmann found is that a person with schizophrenia is much more likely than a person without schizophrenia to have a parent or a sibling with the disorder. Even more compelling, he found that an identical twin of a person with schizophrenia or bipolar disorder is much more likely than a fraternal twin to have the same disorder. Because identical twins share all the same genes and fraternal twins share only half their genes, this finding clearly implicated the identical twins' genes, rather than their shared environment, in the higher incidence of these mental disorders.

Studies of twins show that autism also has a powerful genetic component: when one identical twin has autism, the other identical twin has a 90 percent chance of developing the disorder. A different sibling in that same family, including a fraternal twin, is considerably less likely to develop autism, while an individual in the general population has only a scant chance of developing the disorder (table 1).

TABLE 1. INCIDENCE OF AUTISM AND PSYCHIATRIC DISORDERS IN IDENTICAL TWINS AND SIBLINGS OF AFFECTED INDIVIDUALS

Disorder	*Identical Twins*	*Siblings*	*General Population*
Autism	90%	20%	1–3%
Bipolar disorder	70%	5–10%	1%
Depression	40%	<8%	6–8%
Schizophrenia	50%	10%	1%

We have learned a great deal about the role that genes play in medical disorders by looking at family histories. Based on those histories, it is possible to divide genetic illnesses into two groups: simple and complex (figs. 1.6a and 1.6b).

A simple genetic illness, like Huntington's disease, is caused by a mutation in a single gene. A person who has that mutation will have the disease, and if one identical twin has the disease, they both will. In contrast, vulnerability to a complex genetic disease like bipolar disorder

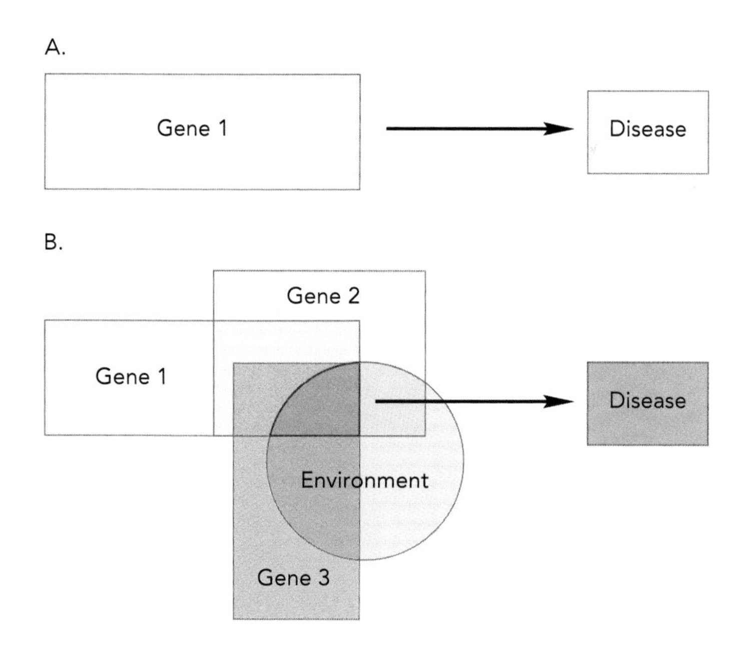

Figure 1.6. A simple genetic illness can involve the mutation of a single gene (A), whereas a complex genetic illness may involve several genes as well as environmental factors (B).

or depression is dependent upon the interaction of several genes with one another and with the environment. We can tell that bipolar disorder is complex because we know that if one identical twin develops the disorder, the other may not. This means that environmental factors must play a key role. When both genes and environment are involved, it is usually easier to find *candidate genes* first, by carrying out large-scale studies to determine which genes correlate with depression and which correlate with mania, and then try to sort out the environmental contribution.

BRAIN IMAGING

Until the 1970s, clinicians had limited tools to examine the living brain: X-rays, which reveal the bony structure of the skull but nothing of the brain itself; angiography, which reveals the blood supply in the brain; and pneumoencephalography, which reveals the ventricles of the brain (the hollow spaces filled with cerebrospinal fluid). Using these crude radiological methods in addition to autopsy, brain scientists for years examined depressed and schizophrenic people but could detect no damage to the brain. In the 1970s, however, two categories of imaging that would dramatically change our understanding of the brain began to emerge: structural imaging and functional imaging.

Structural imaging looks at the anatomy of the brain. Computed tomography (CT) combines a series of X-ray images taken from different angles into a cross-sectional picture. These scans are used to contrast the density of different parts of the brain, such as the bundles of axons that make up the white matter and the cell bodies and dendrites of neurons that make up the cerebral cortex, or gray matter.

Magnetic resonance imaging (MRI) makes use of a very different technique: it contrasts the response of various tissues to applied magnetic fields. The resulting picture provides more-detailed information than computed tomography. For example, MRI has revealed that in people with schizophrenia, the lateral ventricles of the brain are enlarged, the cerebral cortex is thinner, and the hippocampus is smaller.

Functional imaging goes one step further, introducing the dimension of time. Functional imaging enables scientists to observe activity in the brain of a person who is carrying out a cognitive task, such as looking at a work of art, hearing, thinking, or remembering. Functional magnetic resonance imaging (fMRI) works by detecting changes in the concentration of oxygen in red blood cells. When an area of the brain becomes more active, it consumes more oxygen; to meet the demand for more oxygen, blood flow to the area increases. Thus, scientists can use fMRI to create maps showing which parts of the brain are active during a variety of mental tasks.

Functional imaging evolved from studies pioneered by Seymour Kety and his colleagues, who in 1945 developed the first effective way to measure blood flow in the living brain. In a series of classic studies, they measured blood flow in the brains of people who were awake and people who were asleep, thereby establishing the basis for subsequent studies using functional imaging. Marcus Raichle, a pioneer of brain imaging, has noted that the impact of Kety's studies on our understanding of the circulation and metabolism of the human brain cannot be overestimated.

Kety then proceeded to study normal and disordered brain function. He found that overall blood flow in the brain is *not* altered in a surprising range of conditions, from being deeply asleep to being fully awake, from doing mental arithmetic to being mentally disorganized as a result of schizophrenia. This led him to suspect that measuring blood flow in the entire brain doesn't capture important changes that might be taking place in specific regions of the brain. He therefore decided to search for ways of measuring regional blood flow.

In 1955, together with Louis Sokoloff, Lewis Rowland, Walter Freygang, and William Landau, Kety devised a method of visualizing local blood flow in twenty-eight different regions of the cat brain.[2] The group made the remarkable discovery that visual stimulation increases blood flow *only* to the components of the visual system, including the visual cortex, the region of the cerebral cortex that is dedicated to processing visual information. This was the first evidence that changes in blood flow are directly related to brain activity and, presumably, to brain metabolism. In 1977 Sokoloff developed a technique for measuring regional metabolic activity and used that technique to chart where specific functions are located in the brain, thereby providing an independent way for researchers to localize function in the brain.[3]

Sokoloff's discovery laid the foundation for positron emission tomography (PET) and single-photon emission computed tomography (SPECT), the imaging methods that made it possible to visualize brain function in thinking human beings. PET advanced scientists' understanding of the chemistry of brain processes by enabling them to label specific neurotransmitters used by different classes of nerve cells as well as the receptors on target cells that those transmitters act upon.

Structural and functional imaging techniques have given scientists a new way to look at the brain. They can now see which regions of the brain—and sometimes even which neural circuits within those regions—are not working properly.

This information is essential, because the modern view of psychiatric disorders is that they also are disorders of neural circuits.

ANIMAL MODELS

An animal model of a disorder can be engineered in two ways. One, as we have seen, is by identifying the genes in an animal that are equivalent to the human genes thought to contribute to a disorder, altering those animal genes, and then observing the effects on the animal. The second is by inserting a human gene into an animal's genome to see whether it produces the same effects in the animal as in people.

Animal models such as worms, flies, and mice are critical to our understanding of brain disorders. These models have given us insights into the neural circuit of fear that underlies stress, a major contributor to several psychiatric disorders. Animal models of autism have enabled scientists to observe how the expression of human genes that contribute to the disorder alter the animals' social behavior in various contexts.

Mice are the preeminent animal species for modeling brain disorders. Mouse models have given scientists important insights into how rare structural mutations in genes lead to abnormal brain activity in autism and schizophrenia. Moreover, genetically modified mice are proving to be extremely valuable for studying the cognitive deficits of schizophrenia. They can even be used to model environmental risk factors: scientists can expose mice in utero to risks such as maternal stress or activation of their mother's immune system (as might occur when a mother contracts an infection) to determine how such factors affect brain development and function. Animal models make possible controlled experiments that reveal connections between genes, the brain, the environment, and behavior.

NARROWING THE DIVIDE BETWEEN PSYCHIATRIC AND NEUROLOGICAL DISORDERS

Understanding the biological underpinnings of neurological disorders has greatly enriched our understanding of normal brain function—of how the brain gives rise to the mind. We have learned about language from Broca's and Wernicke's aphasias, about memory from Alzheimer's disease, about creativity from frontotemporal dementia, about movement from Parkinson's disease, and about the link between thought and action from spinal cord injuries.

Studies are beginning to show that some diseases that produce different symptoms come about in the same way; that is, they share a common molecular mechanism. For example, Alzheimer's disease, which primarily affects memory; Parkinson's disease, which primarily affects movement; and Huntington's disease, which affects movement, mood, and cognition, are all thought to involve faulty protein folding, as we shall see in later chapters. The three disorders produce strikingly different symptoms because the abnormal folding affects different proteins and different regions of the brain. We will undoubtedly discover common mechanisms in other diseases as well.

Presumably, every psychiatric illness arises when some parts of the brain's neural circuitry—some neurons and the circuits to which they belong—are hyperactive, inactive, or unable to communicate effectively. We don't know whether those dysfunctions stem from microscopic injuries that we can't see when we look at the brain, from critical changes in synaptic connections, or from faulty wiring of the brain during development. But we do know that all psychiatric disturbances result from specific changes in how neurons and synapses function, and we also know that insofar as psychotherapy works, it works by acting on brain functions, creating physical changes in the brain.

Thus, we now know that psychiatric illnesses, like neurological disorders, arise from abnormalities in the brain.

How are psychiatric and neurological disorders different? At the moment, the most obvious difference is the symptoms that patients experience. Neurological disorders tend to produce unusual behavior, or fragmentation of behavior into component parts, such as unusual movements of a person's head or arms, or loss of motor control. By contrast, the major psychiatric disorders are often characterized by exaggerations of everyday behavior. We all feel despondent occasionally, but this feeling is dramatically amplified in depression. We all experience euphoria when things go well, but that feeling goes into overdrive in the manic phase of bipolar disorder. Normal fear and pleasure seeking can spiral into severe anxiety states and addiction. Even certain hallucinations and delusions from schizophrenia bear some resemblance to events that occur in our dreams.

Both neurological and psychiatric disorders may involve reduced function. For example, as there is a loss of control over movement in Parkinson's disease, so there is a loss of memory in Alzheimer's disease, a loss of the ability to process social cues in autism, and reduced cognitive skills in schizophrenia.

A second apparent difference is in how readily we can see actual physical damage to the brain. Damage resulting from neurological disorders, as we have learned, is often clearly visible at autopsy or in structural imaging. Damage resulting from psychiatric disorders is often less obvious, but as imaging techniques improve in resolution, we are beginning to detect changes resulting from these disorders. For example, as mentioned previously, we can now identify three structural changes in the brains of people with schizophrenia: enlarged ventricles, thinner cortex, and a smaller hippocampus. Thanks to improvements in functional brain imaging, we can now observe certain changes in brain activity that are characteristic of depression and other psychiatric disorders. Finally, as techniques for detecting even subtler damage to nerve cells become available, we should be able to find such damage in the brains of all people with psychiatric disorders.

The third apparent difference is location. Because of neurology's traditional emphasis on anatomy, we know a great deal more about the neural circuitry of neurological disorders than of psychiatric disorders. In addition, the underlying neural circuitry of psychiatric disorders is more complex than that of neurological disorders. Scientists have only recently begun to explore the brain regions involved in thought, planning, and motivation, the mental processes that are disordered in schizophrenia and in mood and emotional states such as depression.

Some psychiatric disorders, at least, do not appear to involve permanent structural changes in the brain and therefore are more likely to be reversible than disorders stemming from obvious physical damage. For example, scientists have found that increased activity in a particular area of the brain is reversed in successful treatment of depression. That said, newer treatments may eventually reverse even the physical damage caused by neurological disorders, as is now being done in some people with multiple sclerosis.

As research into the brain and mind advances, it appears increasingly likely that there are actually no profound differences between neurological and psychiatric illnesses and that as we understand them better, more and more similarities will emerge. This convergence will contribute to the new, scientific humanism, offering a chance to see how our individual experiences and behavior are rooted in the interaction of genes and environment that shapes our brains.

2

OUR INTENSELY SOCIAL NATURE: THE AUTISM SPECTRUM

We are by nature intensely social beings. Our success in adapting to the natural world over the course of evolution has resulted in large part from our ability to form social networks. More than any other species, we depend on one another for companionship and survival. As a result, we cannot develop normally in isolation. Children are innately prepared to construe the world they will encounter as adults, but they can learn the critical skills they will need, such as language, only from other people. Sensory or social deprivation early in life can impair the structure of the brain. Similarly, we need social interaction to keep the brain healthy in old age.

We have learned a great deal about the nature and importance of our social brain—the regions and processes that are specialized for interaction with other people—by studying autism, a complex disorder in which the social brain does not develop normally. Autism appears during a critical period of development early in life, before the age of three. Because autistic children are not able to develop social and communication skills spontaneously, they withdraw into an inner world and do not interact socially with others.

Autism includes a spectrum of disorders ranging from mild to severe, all of which are characterized by difficulty connecting with others. People with autism have an impaired ability to engage in social interactions and communications, both verbal and nonverbal; in addition, their interests are restricted. Such barriers to interaction with others profoundly affect social behavior.

This chapter explores what autism has taught us about our social brain, including our ability to read the mental and emotional states of others. It describes cognitive psychology's contribution to our understanding of autism and the insights into the neural circuitry of the social brain that autism studies have given us. Scientists have yet to discover the causes of autism, but genes appear to play a leading role. Remarkable new advances in genetics show how mutations in certain genes disturb key biological processes during development, resulting in autism spectrum disorders. Finally, we will touch on what we have learned from social behavior in animals.

AUTISM AND THE SOCIAL BRAIN

Based on their studies of chimpanzees, David Premack and Guy Woodruff of the University of Pennsylvania proposed in 1978 that each of us has a *theory of mind*—that is, we attribute mental states to ourselves and to others.[1] Each of us has the ability to appreciate that other people have a mind of their own, that they have their own beliefs, aspirations, desires, and intentions. This innate understanding is different from a shared emotion. A very young child will smile when you smile or frown when you frown. But realizing that the person you're looking at may be thinking about something different from what you're thinking about is a profound skill that arises only later in normal development, around the age of three or four.

Our ability to attribute mental states to others enables us to predict their behavior, a critical skill for social learning and interaction. When you and I talk, for example, I have a sense of where you are going in the conversation and you can sense where I am going. If you are joking with me, I will not interpret you literally and will predict different behavior from you than I would if I had the sense that you were speaking seriously. In 1985 Uta Frith, Simon Baron-Cohen, and Alan Leslie at University College London applied the concept of theory of mind to people with autism.[2] Frith (fig. 2.1) describes how this came about:

Figure 2.1. Uta Frith

How does the mind work? What does it mean to say the mind is created by the brain? Since my student days in experimental psychology I have been passionately interested in these sorts of questions. Pathology was the obvious way to get at possible answers, and I trained to be a clinical psychologist at the Institute of Psychiatry in London. Here I met autistic children for the first time. They were completely fascinating. I wanted to find out what makes them behave so strangely with other people, and what made them so totally untouched by the kind of everyday communication we take for granted. I still want to find out! Because even a lifetime of research is not enough to get to the bottom of the enigma that is autism. . . .

I wanted to know why autistic individuals, even when they had good language, were so difficult to involve in a conversation. The concept of "theory of mind" was just then being developed by bringing together studies from animal behavior, philosophy, and developmental psychology. It seemed to me and my then-colleagues Alan Leslie and Simon Baron-Cohen [to be] of extreme interest to autism, possibly the key to their social impairments. And so it proved to be.

> We started systematic behavioral experiments in the 1980s and showed that autistic individuals indeed do not show spontaneous "mentalizing." That is, they do not automatically attribute psychological motives or mental states to others to explain their behavior. As soon as neuroimaging methods became available we scanned autistic adults and revealed the brain's mentalizing system. This work is still ongoing.[3]

Research into autism has taught us a great deal about social behavior and the biology of social interactions and empathy. Some social interactions, for example, occur through biological motion—walking toward another person, reaching a hand out in greeting. In 2008 Kevin Pelphrey of Yale University, then at Carnegie Mellon University, discovered that autistic children have difficulty distinguishing biological motion.[4] In an experiment with autistic and non-autistic (neurotypical) children, he monitored two regions of the brain while the children were looking at biological or non-biological motion. One brain region was a small visual area known as MT or V5 (MT/V5), which is sensitive to any motion; the other was the superior temporal sulcus, which in neurotypical adults responds more strongly to biological motion. The biological motion Pelphrey showed the children was a person or a human-like robot walking; the non-biological motion was a disjointed mechanical figure or a grandfather clock. In both groups of children, the motion-sensitive MT/V5 region of the brain responded about equally to the two kinds of motion. But in typically developing children the superior temporal sulcus responded more strongly to biological motion. In autistic children the same brain area did not register any difference between the two kinds of motion (fig. 2.2).

The ability to identify and integrate biological action with the context in which it occurs—for example, to integrate our observation that a person is reaching for a glass of water with our surmise that that person is thirsty—enables us to recognize intention, which is critical to a theory of mind. Thus, one of the reasons people with autism have difficulty with social interactions is that they have limited capacity to read socially meaningful biological actions such as reaching to shake hands.

People with autism have a similar difficulty reading faces. When autistic people look at another person, they avoid the eyes and instead tend

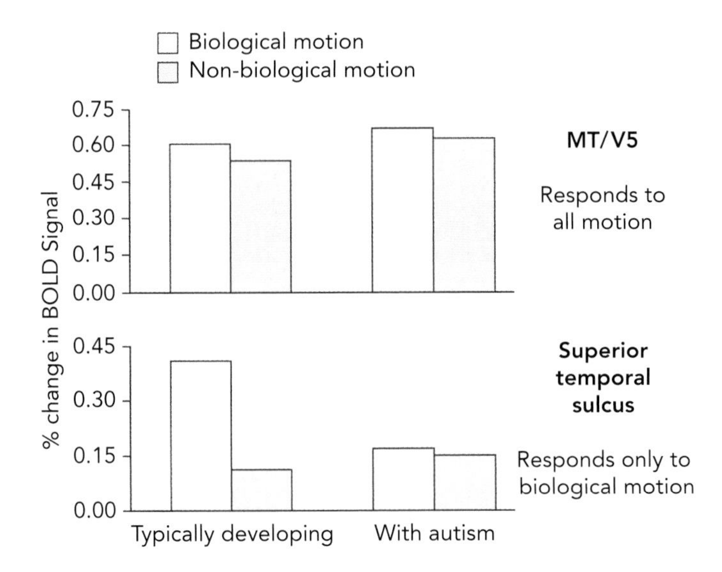

Figure 2.2. Responses to biological and non-biological motion in two regions of the brain in typically developing children and children with autism. MT/V5 is a region of the occipital lobe.

to look at the mouth (fig. 2.3). Neurotypical people do the reverse: they look primarily at the eyes. Why? Because a person's gaze—where he or she is looking—gives us important clues about what that person desires, intends, or believes. The words "desire," "intend," and "believe" describe mental states. States of mind are not actually open to direct observation, but most of us behave as if we can directly observe another person's mental states, as if we can read minds.

Take the marvelous painting *The Cheat with the Ace of Diamonds*, by Georges de La Tour (fig. 2.4). What do you see when you look at it? You are probably attracted to the strange gaze of the lady sitting down. She is obviously communicating with the woman standing to her right. The standing woman has seen the cards in the hands of the player on the left. This player is a cheat: you can see that he is hiding the ace of diamonds behind his back. The player at the right is a rich young man who will be cheated out of the heap of gold coins in front of him.

How can we interpret this scene, painted almost four centuries ago, so confidently? How can the painter rely on us to put together all the clues he has given—the gaze, the pointing finger, the hidden card—and arrive at the correct interpretation? Our uncanny skill derives from our

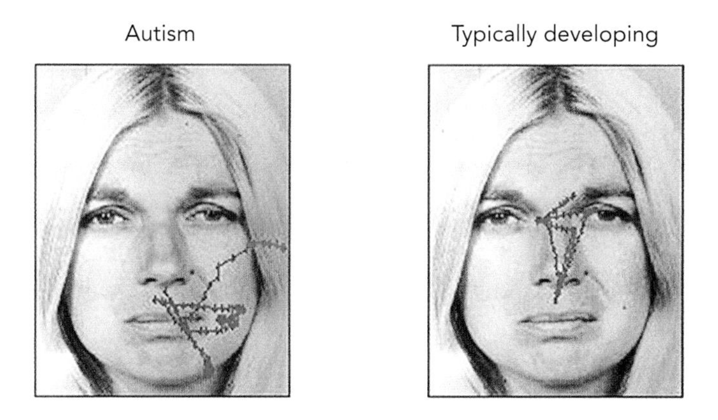

Figure 2.3. Eye movement patterns in an autistic versus a neurotypical person

Figure 2.4. Georges de La Tour, *The Cheat with the Ace of Diamonds*, ca. 1635, Louvre, Paris

ability to formulate a theory of mind. We use it all the time to explain and predict other people's behavior.

A major disturbance in autism occurs in the connections between gaze and intention. Although we still have a long way to go to understand the biological causes of autism—the genes, synapses, and neural circuitry that are altered—we do know quite a bit about the cognitive psychology of autism and, through it, about the cognitive systems in our brain that are responsible for the theory of mind.

THE NEURAL CIRCUITRY OF THE SOCIAL BRAIN

In 1990 Leslie Brothers of the UCLA School of Medicine took advantage of the insights into theory of mind that had been derived from studies of autism to propose a theory of social interaction.[5] She argued that social interaction requires a network of interconnected brain regions that process social information and together give rise to a theory of mind; she coined the term *social brain* to describe this network. The regions include the inferior temporal cortex (involved in face recognition), the amygdala (emotion), the superior temporal sulcus (biological motion), the mirror neuron system (empathy), and the areas in the temporal-parietal junction involved in theory of mind (figs. 2.5 and 2.6).

Brain science is only now beginning to decipher how the regions of the social brain identified by cognitive psychology are connected and how they interact to affect behavior. Stephen Gotts and his colleagues at the National Institute of Mental Health have used functional brain imaging to confirm that the neural circuit of the social brain is indeed disrupted in people with autism spectrum disorders. Specifically, disrupted connections occur in three regions of the social brain: those involved in the emotional aspects of social behavior, those involved in language and communication, and those involved in the interplay between visual perception and movement. Normally, patterns of activity in these three regions are coordinated with one another, but in people with autism they are not. Rather, they are out of sync with one another and with the rest of the social brain.[6]

Of particular interest are anatomical findings regarding the timing of brain growth and development in autistic children. Before the age of two, the circumference of an autistic child's head is often larger than that of a typically developing child. In addition, some regions of an autistic child's brain may develop prematurely during the first years of life, particularly the frontal lobe, which is involved in attention and in decision making, and the amygdala, which is involved in emotions.[7]

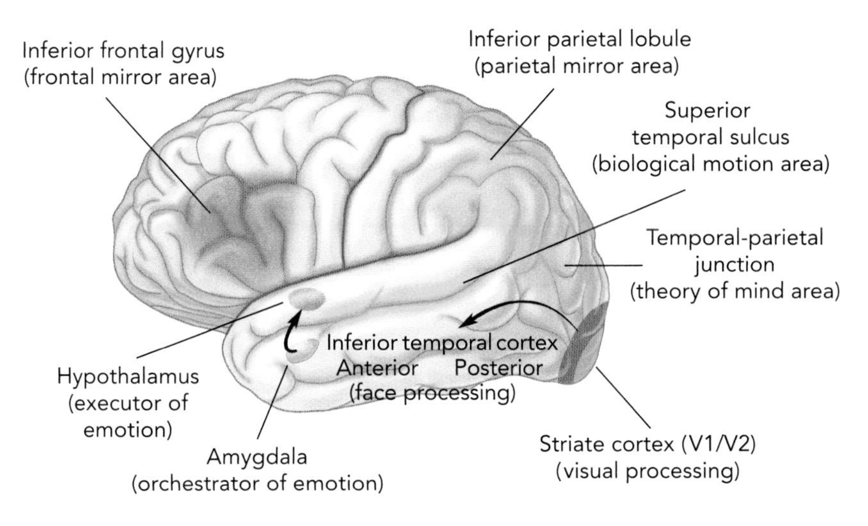

Figure 2.5. The network of regions that makes up our social brain

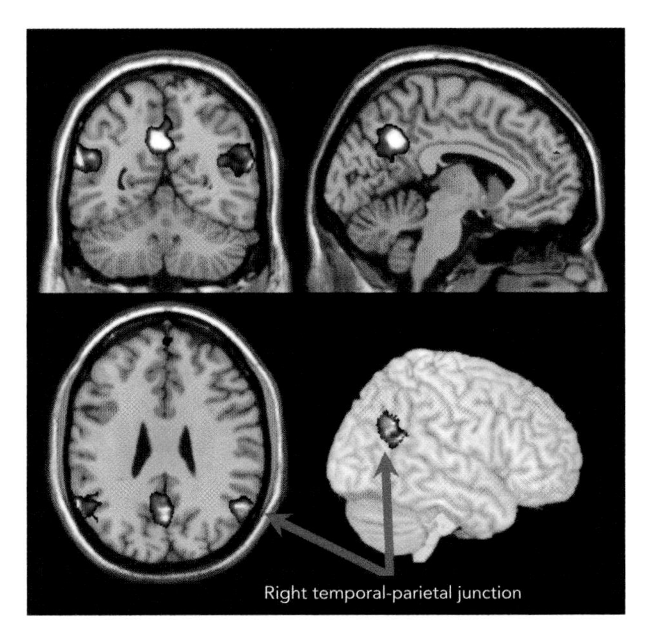

Figure 2.6. Theory of mind: the neural mechanism in the temporal-parietal junction that is recruited when thinking about someone else's thoughts, beliefs, or desires

This is important because when one or more regions of the brain develop out of sequence, they may seriously disturb the patterns of growth in other regions of the brain to which they connect.

THE DISCOVERY OF AUTISM

Autism was recognized as a separate disorder in the early 1940s by two scientists who had no contact with each other: Leo Kanner, working in the United States, and Hans Asperger, working in Austria. Until then, children with the disorder had been diagnosed as mentally retarded or as having behavioral disorders.

Remarkably, Kanner and Asperger not only came up with similar descriptions of what they were studying, they even gave the disorder the same name, *autism*. The word had been introduced into the clinical literature by Eugen Bleuler, the great Swiss psychiatrist who coined the term *schizophrenia*. Bleuler used "autistic" to refer to a particular group of symptoms that characterize schizophrenia: social awkwardness, aloofness, and an essentially solitary life.

Kanner was born in Austria and educated in Berlin. He moved to the United States in 1924 and took a position at the state mental hospital in Yankton, South Dakota. From there he went to The Johns Hopkins University, where he founded the Children's Psychiatric Clinic in 1930. In 1943 he wrote his classic paper "Autistic Disturbances of Affective Contact," in which he described eleven children.[8] One of them, Donald, was happiest when he was alone. Kanner prefaced his own observations of Donald with a description written by the boy's father: "'He seems almost to draw into his shell and live within himself . . . oblivious to everything about him.' In his second year, he 'developed a mania for spinning blocks and pans and other round objects.' . . . He . . . developed the habit of shaking his head from side to side." Based on his analysis of Donald and the ten other children, Kanner presented a vivid picture of the three important features of classic autism in childhood: (1) profound aloneness, a strong preference for being by oneself; (2) a desire for things to be the same, not to change; and (3) islets of creative ability.

Asperger was born just outside Vienna. He received his medical degree from the University of Vienna and worked at the university's pediatric clinic. Asperger realized that autism does not take the same form in all people with the disorder. It covers a wide spectrum, from people who are below average in some intellectual activities and have great difficulty with language, to those who are very bright and have no difficulty with language. Moreover, he found that autism persists and is evident in adults as well as children.

The children Asperger saw were on the mild end of the autism spectrum. Some of them functioned on a very high intellectual level; for example, Elfriede Jelinek, who was awarded a Nobel Prize in Literature, was a patient of Asperger's. Until quite recently, high-functioning autistic children and adults were diagnosed as having Asperger's syndrome. Today, Asperger's syndrome is generally considered part of the autism spectrum.

LIVING WITH AUTISM

Being the parent of an autistic child is difficult. Alison Singer, the president of the Autism Science Foundation, who has a daughter with autism, describes it as "a challenge and a struggle every day. . . . It's financially exhausting. It's emotionally exhausting. It's 24/7 taking care of someone who can't really communicate, with whom I can't really communicate. I have to surmise most of the time what she's trying to say."

As Singer explains:

> Living with an autistic child is really about trying every day to find a balance between loving your child for exactly who she is and constantly pushing for more. And by "more" I mean more language, more social interaction, more restaurants or other places in the community where she can go without having a meltdown.
>
> My daughter exhibited a lot of the typical early warning signs of autism. When she was a baby, she never babbled. She never had social gestures. She never waved bye-bye. She never shook her head yes or no. She had tantrums that were off the charts. She really struggled to make eye contact when I would take her to the playground or to play groups. She never showed any interest in the other children. She did have some words, but all of them were

> words she had heard from books or videos and she didn't use them in a meaningful way, to communicate. She just repeated them over and over. And she played with toys in very unusual ways. She would sort them by color, line them up by size; she never really played with them in the way that the toy manufacturers intended them to be played with. Don't be fooled into thinking your child is using toys in some sort of "creative way." Really, toys should be used the way the manufacturer intended.
>
> As she's gotten older—she is nineteen and a half now—some of those symptoms have become more ingrained, more entrenched, but in other ways she has improved. Autism is a developmental disorder and as they get older, most kids show improvement. Some is the result of intensive therapy, and some is just maturity.[9]

In the 1960s Bruno Bettelheim, a Viennese-born psychologist who specialized in emotionally disturbed children, popularized the unfortunate term *refrigerator mother* to explain the origins of autism. Bettelheim argued that autism did not have a biological basis but was the result of a mother withholding affection for a child she did not want. Bettelheim's theories about autism, which caused great pain to many parents, have been completely discredited.

Singer is grateful that research has revealed the biological basis of autism:

> At least now we no longer have to struggle with the idea that autism is the result of bad parenting and that parents of children with autism were too cold to bond properly with their child and that that is what caused the child to retreat into her own world. The parents of children with autism love their children more than you could ever understand. We do everything, everything to help them gain skills and participate in community activities.
>
> When my brother was diagnosed with autism in the 1960s, my mother was told that she was a "refrigerator mother," too cold to bond with my brother, and that his autism was her fault. The doctor said she should try harder with her next child. Thank goodness those days are behind us. We know that autism is a genetic

> disorder, and we are learning more and more every day about the genes that cause autism. Important research is being done now to understand the causes of autism and to develop better treatments for people with autism.[10]

Once it became clear that autism has a biological basis, scientists could begin to refine our understanding of the disorder. They have found, for example, that the social interactions of people with less severe autism are guided by actual behavior, not by the intentions hidden in behavior. This makes it hard for those autistic people to spot ulterior motives and manipulation, a bit like the naïve young man playing cards in the painting in figure 2.4. Severely impaired autistic people are inherently straightforward and honest: they feel no pressure to conform to other people's ideas and beliefs. Autistic people who function at a high level in social situations do feel pressure to conform, but they have no innate sense of how to do it. This lack of an internal social compass contributes to the depression and anxiety that children who are on the mild end of the autism spectrum often experience.

Learning about mental states such as believing, desiring, and intending does not eliminate problems with social communication; it just alleviates them. Even the most able, highly adapted people on the autism spectrum have some difficulty deciphering and interpreting mental states. They need time to do it. Written communications, such as e-mail, are easier than face-to-face interactions. Nevertheless, it would be a mistake to underestimate the stress and anxiety that most individuals on the autism spectrum experience in trying to fit into a world of neurotypical people.

Erin McKinney, who has autism, describes how she experiences the stresses caused by the disorder:

> Autism makes my life loud. That's the best adjective I have found. Everything is amplified. I don't mean this only in terms of my sense of hearing, although that is one part of it. I feel loudly. A light touch feels not so light. A bright light feels brighter. A soft buzzing from a light feels thunderous. Instead of happy, I feel overwhelmed. Instead of sad, I feel overwhelmed. The general perception is that autistic people don't feel empathy. I, along

> with most individuals on the spectrum, find the reverse is true. . . . Autism makes my life stressful. When everything is louder, situations tend to be a little more stressful.[11]

When she was first diagnosed with autism, McKinney says, she felt "very conflicted." But soon, she felt thankful for having received the diagnosis and began the difficult, continuing work of coming to terms with it:

> I live my life constantly on edge. And sometimes, I fall off the edge and a meltdown occurs. And that's OK. Well, maybe it isn't OK. But it has to be. I don't have a choice. . . . I have to keep going. I work hard to notice when I see myself trending toward a meltdown, so I can change course. It has taken a lot of work for me to get to this point of self-awareness, but it still doesn't work all the time.
>
> . . . I do the same thing the same way every time. I count lots of things, notice things that most think are unimportant, and stress over tiny imperfections. I get thoughts stuck in my head, over and over and over again. Phrases, images, memories, patterns. These can all become overwhelming. I use them to my advantage as much as I can. I think this is part of why I am good at my job. And I am very good at my job. I notice the little things, the nuances that others tend to overlook. I find the pattern and I find it quickly.[12]

In reflecting on her life, McKinney concludes:

> There is no doubt that autism makes my life difficult, but it also makes my life beautiful. When everything is more intense, then the everyday, the mundane, the typical, the normal—those things become outstanding. I can't speak for you or anyone else, on the spectrum or not. Our experiences are all unique. Regardless, I do believe that it is important to find the beautiful. Recognize that there is bad, there is ugly, there is disrespect, there is ignorance, and there are meltdowns. Those things are inevitable. But there is also good.[13]

About 10 percent of people with autism have low IQ scores, but many have special talents for writing poetry, learning foreign languages, performing music, drawing and painting, calculating, or knowing the day of the week for any date in the calendar. In *Bright Splinters of the Mind*, a book about her research with autistic people, the experimental psychologist Beate Hermelin noted that autism researchers are continually fascinated with the remarkable talents displayed by these autistic savants.[14] One of the best-known autistic savants is Nadia. When she was a young girl, between the ages of four and seven, Nadia made a number of drawings that were generally admired, even by professionals, and that compared in beauty to the cave paintings from thirty thousand years ago. We will discuss the creative capabilities of people with autism more fully in chapter 6.

THE ROLE OF GENES IN AUTISM

Scientists have known for years that genes play an extremely important role in autism. Studies of identical twins, who have the same genetic makeup, show that if one twin has autism, the chances are up to 90 percent that the other twin will be autistic. No other developmental disorder has as high a concordance between identical twins.

This startling finding has convinced many scientists that the quickest route to understanding the brain mechanisms involved in autism is to focus on the genetics of the disorder. Once scientists have scoped out the genetic landscape and understand what the risk factors are, they will be in a much better position to figure out where in the brain these genes are working. However, autism is not a simple one-gene, one-disease disorder. Many genes are likely to contribute to the risk of autism.

At the same time, we cannot rule out environmental factors, because all behavior is shaped by the interplay between genes and the environment. Even a mutation in a single gene that invariably causes a disease can be influenced strongly by the environment. Take phenylketonuria (PKU), a simple metabolic disease that babies are routinely tested for at birth. This rare genetic disorder affects one person in fifteen thousand and can result in severe impairment of cognitive functioning. People with the disease have two abnormal copies of the gene that is ultimately responsible for breaking down the amino acid phenylalanine, a component of protein in the foods we eat. (People with just one defective copy of the gene don't develop PKU.) If the body can't break down phenylalanine, it builds up in the blood, leading to the production of a toxic substance that interferes with normal brain development. Fortunately, mental retardation can be completely prevented by a simple, amazingly effective environmental intervention—restricting the amount of protein that people at risk of PKU eat.

Dramatic advances in our ability to study DNA at high resolution and in many people have begun to give scientists a clearer view of the genetic landscape. Those technological advances have transformed our understanding of how DNA varies among people and how some variations lead to disorders like those on the autism spectrum. Specifically, they have revealed two previously unknown types of genetic aberrations: *copy number variations* and *de novo mutations*. Both contribute to autism as well as to schizophrenia and other complex disorders produced by mutations in more than one gene.

COPY NUMBER VARIATIONS

We all have slight differences in the nucleotide sequences of our genes. (Nucleotides, as we learned in chapter 1, are the molecules that make up DNA.) These slight differences are called *single-nucleotide variations* (fig. 2.7). About a decade ago, scientists discovered that we may also have major differences in the structure of our chromosomes. These rare structural differences are known as copy number variations (fig. 2.8). We may be missing a small bit of DNA from a chromosome (a copy number deletion), or we may have an extra bit of DNA in a chromosome (a copy number duplication). Copy number variations may decrease or increase the number of genes on a chromosome by twenty to thirty genes, but in either case they heighten the risk of autism spectrum disorders.

Copy number variations have given us a better understanding of the specific genes involved in autism, which in turn has afforded us a much better view of the molecular basis of social behavior. A case in point is copy number variations on chromosome 7. Matthew State, now at the University of California, San Francisco, has found that having an extra copy of one segment of chromosome 7 puts people at much greater risk of developing a disorder on the autism spectrum. When that same brain region is lost, however, the result is Williams syndrome.[15]

Williams syndrome is virtually the reverse of autism. Children with this genetic disorder are extremely social (fig. 2.9). They have a strong,

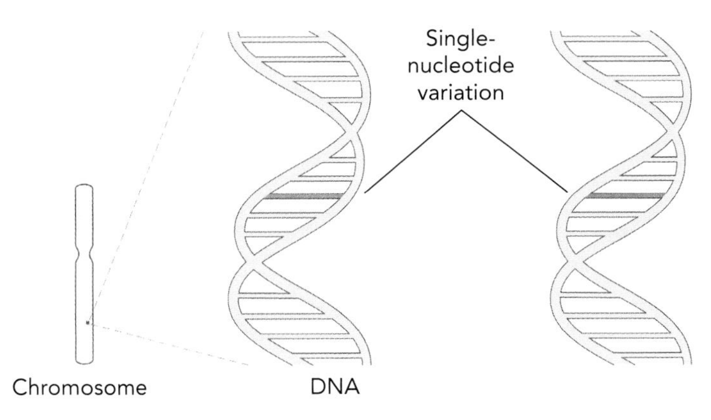

Figure 2.7. Single-nucleotide variation

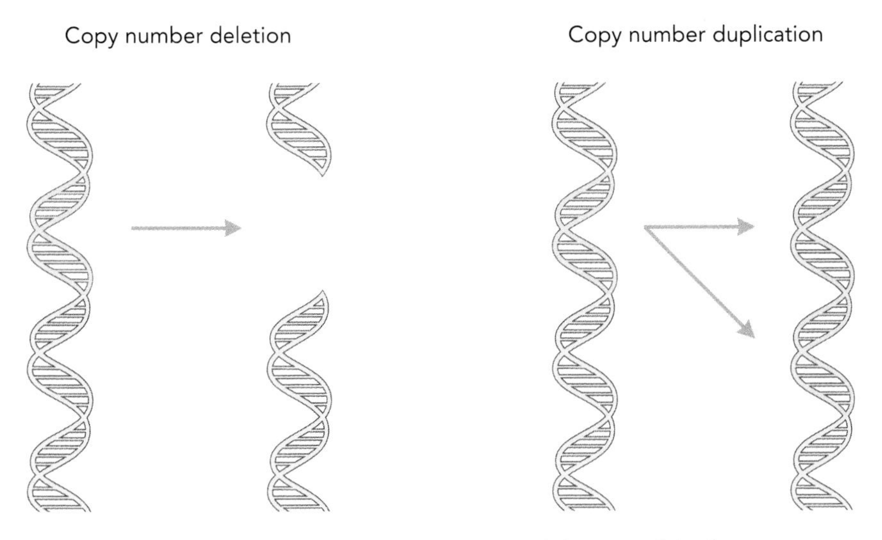

Figure 2.8. Copy number variations: DNA deletion and duplication

almost irrepressible desire to speak and communicate. They are very friendly and trusting, even of strangers. Moreover, whereas some children with autism have strong drawing skills, children with Williams syndrome tend to be musical. In fact, children with Williams syndrome have difficulty constructing visuospatial relationships, which may account for their inability to draw well. Unlike children with autism, children with Williams syndrome have good language skills and do well with face recognition; they have no difficulty reading the emotions of others and gauging their intentions.

Thomas Insel, formerly the director of the National Institute of Mental Health, argues that the contrast between autism and Williams syndrome suggests that our brain uses specific networks for specific types of functions, such as social interaction. Deficits in the functioning of the

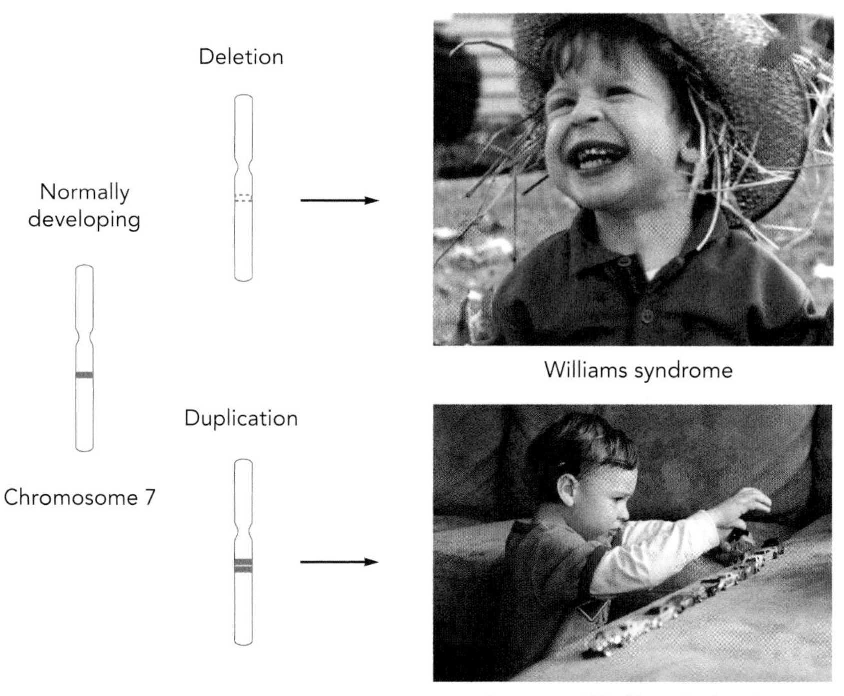

Figure 2.9. Copy number variations: deletion of a particular segment on chromosome 7 causes Williams syndrome, whereas duplication of that segment increases the likelihood of developing an autism spectrum disorder.

social network may lead the brain to compensate by developing expertise in a non-social network, resulting in the kinds of unusual abilities we see in autistic savants.[16]

The fact that this single segment, containing about twenty-five of the twenty-one thousand or so genes in our genome, could have such a profound influence on complex social behavior is astonishing. This kind of discovery gives scientists something very specific to pursue and should open important new avenues in developing treatments.

DE NOVO MUTATIONS

The second genetic aberration revealed by advances in technology is the recent discovery that not all mutations are present in the genomes of our parents. Some mutations arise spontaneously in the sperm of adult men. These rare spontaneous mutations are called de novo, or new, mutations, and a father can transmit them to his children. Four nearly simultaneous studies by scientists at Yale, the University of Washington, the Broad Institute at the Massachusetts Institute of Technology, and Cold Spring Harbor Laboratory have found that de novo mutations markedly increase the risk of autism.[17]

Moreover, the number of de novo mutations increases with paternal age. A recent study led by deCODE Genetics, a biotechnology company in Iceland, confirmed this finding using a genome-wide technique in which all of the DNA in a person's genome, not just the portion coding for proteins, is studied.[18] This is important because scientists have recently discovered that the noncoding DNA in our genome, formerly considered "junk," may play a major role in complex diseases by switching genes on and off.

The reason de novo mutations increase with age is that sperm precursor cells divide every fifteen days. This continued division and copying of DNA leads to errors, and the rate of error increases significantly with age. Thus, a father who is twenty years old will have, on average, twenty-five de novo mutations in his sperm, whereas a father who is forty years old will have sixty-five mutations (fig. 2.10). Most of these mutations are harmless, but some are not: de novo mutations are now thought to contribute to at least 10 percent of autism cases. Mothers do not appear to contribute to autism by means of de novo mutations because egg cells, unlike sperm, do not divide and multiply throughout life; they are all generated before a woman is born.

De novo mutations are particularly interesting because the incidence of autism has increased substantially in recent years. A large part of this upswing is probably attributable to the fact that we are now much more aware of the disorder and are better at identifying it than we were fifty years ago. But another part of the explanation is that people are having children at a later age. We now know that older fathers are more prone to de novo mutations in their sperm and are therefore more likely to pass on these mutations—and thus a greater risk of autism—to their children.

We also have evidence that de novo mutations in an older father's sperm contribute to schizophrenia (fig. 2.10) and to bipolar disorder.

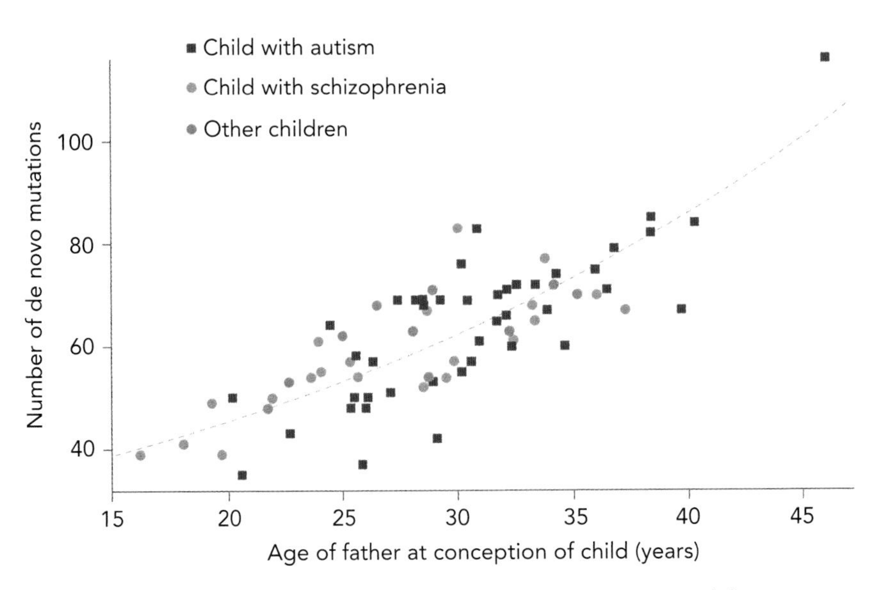

Figure 2.10. Paternal influence: researchers analyzed genetic material from seventy-eight Icelandic children and their parents, including forty-four children with autism. Children of older fathers tended to have a higher number of de novo mutations, which are not present in the genomes of either parent.

(As Bleuler observed a century ago, some of the same social difficulties that characterize autism are shared by people with schizophrenia.) Moreover, we know that schizophrenia and bipolar disorder are not caused by single genes. Thus the assortment of possible genetic culprits responsible for autism seems to be common to these psychiatric disorders as well. We don't know exactly how many genes are capable of contributing to autism, but there are very likely at least fifty, and more likely hundreds, of such genes.

Finally, de novo mutations might account for another interesting feature of autism: the disorder is not dying out. Autistic adults are less likely to have children than neurotypical people, yet the number of children diagnosed with autism spectrum disorders each year has not dropped. De novo mutations in the sperm of fathers who do not have autism could be one reason for the persistence of autism in the population at large.

NEURAL CIRCUITS AS TARGETS FOR MUTATIONS

A recent study revealed that the brains of adolescents with autism have too many synapses.[19] Ordinarily, the excess synapses in our brain—the synapses we don't use—are removed in a process known as *synaptic pruning*, which begins quite early in childhood and peaks in adolescence and early adulthood. The finding of too many synapses indicates that not enough of them were pruned out, resulting in a thicket of neural connections rather than streamlined, efficient neural circuits. Interestingly, while autism involves insufficient synaptic pruning, schizophrenia involves excessive synaptic pruning, as we shall see in chapter 4.

The process of wiring the developing brain is extraordinarily complex and presents a wide-open opportunity for mistakes. Moreover, roughly half of our genes are active in the brain, and the formation of synapses between neurons requires a huge number of proteins to function normally. Proteins, you will recall, are synthesized according to instructions sent out by genes. If mutations in these genes disrupt the composition or operation of normal proteins at the synapse, a cascade of events results: synapses do not function properly, neurons cannot communicate with one another, and the neural circuits they form are disrupted.

The genetic mutations that contribute to autism spectrum disorders may be distributed anywhere on our twenty-three pairs of chromosomes. Regardless of where they are located, these mutations disturb neural circuits in the social brain, and those disturbances end up compromising theory of mind.

Some of the mutations play key roles in the working of synapses. In

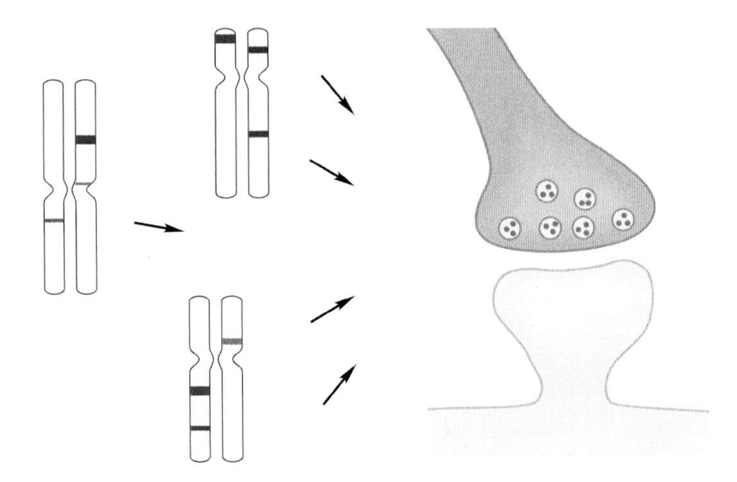

Figure 2.11. Hundreds of genes throughout the genome contribute to synaptic function. A mutation in any one or in a combination of those genes could result in a disorder such as autism. By developing drugs that target the synapse rather than particular genes, we may be able to treat these genetically complex disorders.

fact, de novo mutations occur more frequently in genes that code for synaptic proteins. This fact raises the exciting possibility that autism and other developmental disorders may be amenable to treatment. In other words, we may be able to treat a genetic disorder by fixing faulty synapses (fig. 2.11).

This is a fundamental change in thinking. Rather than being conditions that are set in stone from birth, developmental disorders may prove to be reversible, or at least treatable throughout life.

GENETICS AND SOCIAL BEHAVIOR IN ANIMAL MODELS

Most animals spend at least part of their lives in association with others of their own kind. We acknowledge that reality in the way we talk about them—as schools of fish, flocks of geese, or hives of bees. Clearly, animals recognize one another, communicate with one another, and generate coherent behavior. The naturalist E. O. Wilson noticed that many of the social behaviors of animals are similar, even in animals that are very different. When anyone makes an observation like that in biology, it usually means that the underlying genetics are very ancient and that they are contributing to the same output in many different animals. In fact, almost all of our own genes are present in other animals.

Because both social behavior and genes are conserved through evolution, scientists studying the genetic underpinnings of behavior often turn to simple animals such as the tiny worm *Caenorhabditis elegans* and the fruit fly *Drosophila*. Cori Bargmann, a geneticist at The Rockefeller University, who now heads the Chan Zuckerberg Initiative, studies *C. elegans*, which lives in the soil and eats bacteria. Most members of this species want to spend their time with their fellow worms. Although they sometimes wander off by themselves, they always come back and join the group. This behavior is not about food—food is available everywhere—and it's not about mating. The animals are social; they simply like to associate with one another.

Some of the worms are solitary, however. While feeding, they distribute themselves separately on a lawn of bacteria. The difference between the social and the solitary strains is due to a natural variant in a single gene, which is attributable to a change in a single nucleotide.[20]

Sociability and solitariness can also be traced to a single gene in more-complex animals. While Thomas Insel was at Emory University, he and his colleagues explored the role of the hormone oxytocin in the prairie vole, a ratlike rodent.[21] They found that the hormone stimulates milk production and regulates maternal-infant bonding, as well as other social behavior. To raise their young, male and female prairie voles form enduring pair bonds. These bonds are stimulated by the release of oxytocin in the female vole's brain during mating and the release of a related hormone, vasopressin, in the male's brain. Vasopressin also contributes to paternal behavior.

While male prairie voles form secure pair bonds and help the females raise their offspring, males of a closely related species, the montane vole, breed widely and promiscuously and exhibit no paternal behavior. The difference between the two species is correlated with the number of vasopressin receptors—and hence the amount of vasopressin—in the male brain. Prairie voles have large concentrations of vasopressin in the brain region that deals with pair bonding, whereas montane voles do not. In the two species, variations in oxytocin concentrations in specific areas of the brain account for the differences in pair and parental bonding.[22]

Increasing evidence points to an important role for oxytocin and vasopressin in human pair bonding and child rearing as well. Oxytocin is a peptide hormone that is produced in the hypothalamus and released into the bloodstream by the posterior pituitary gland. Oxytocin regulates milk production by mothers in response to their infants' suckling. Moreover, this hormone enhances positive social interaction by increasing our feelings of relaxation, trust, empathy, and altruism. Sarina Rodrigues at Oregon State University has found that genetic variations in oxytocin production affect empathic behavior: people with less of the hormone in their brain have more difficulty reading faces and feeling distress at other people's suffering.[23]

Other research has suggested that oxytocin may affect our social cognition.[24] When inhaled, the hormone seems to dampen our response to frightening stimuli. It is also thought to enhance positive communication. In some rare cases, nasal inhalation of oxytocin has even improved the social skills of people with autism. Oxytocin increases trust and the willingness to bear risks—essential features for friendship, love, and the organization of family.

As these studies illustrate, some of the same hormones—and therefore the same genes—contribute to social behavior in people and in animals, suggesting that mutations in those genes may contribute to autism spectrum disorders. One can also explore aspects of the biological basis of autism by creating animal models of the disorder. David Sultzer at UCLA and his colleagues, for example, have found a drug that restores normal synaptic pruning in mouse models of autism, thus reducing the animals' autistic-like behavior.[25] Clearly, genetic studies in animals, as well as in people, can be tremendously valuable in understanding how a system as complex as our social brain can go awry.

LOOKING AHEAD

Scientists have gone from largely searching in the dark to having the tools at hand to make major strides in the genetics of autism. With the new technologies that have emerged in the last few years—being able to sequence entire genomes rapidly and rather cheaply, for one—scientists should be able to identify more of the critical autism genes in the future.

Four points stand out in this search. First, hundreds of different genes are capable of contributing to autism spectrum disorders—not necessarily hundreds of genes within a single person, but hundreds of genes across the population. Second, while a mutation in a single gene is responsible for some disorders, such as Huntington's disease, single mutations do not cause most other brain disorders, including autism, depression, bipolar disorder, and schizophrenia. Third, if we can find the genes that contribute to autism, we will be well on the road to understanding what is going wrong at the cellular and molecular levels. Some of the earliest genetic discoveries in autism have pointed to malfunctioning synapses.

Finally, as we identify the genes that contribute to autism, we will gain a better understanding of the genes and the neural pathways that give rise to the social brain—the genes that make us the social beings we are. Moreover, we will learn how genetic predisposition interacts with environmental factors to give rise to specific disorders.

3

EMOTIONS AND THE INTEGRITY OF THE SELF: DEPRESSION AND BIPOLAR DISORDER

We all experience emotional states. In fact, our language overflows with colorful descriptions of how we feel: I got up on the wrong side of the bed. He's singing the blues. She's over the moon about her new job. In these contexts, we are describing emotion as a temporary state of mind that comes and goes. Such shifts of emotion are entirely normal—and desirable. Emotional awareness is vital to staying alive and to negotiating the complexities of human social existence.

A person's emotional state is usually transient and occurs in response to a specific stimulus in the environment. When a particular emotional state becomes fixed and extended in time, we call it a mood. Think of emotion as the daily weather and mood as the prevailing climate. Just as climates range widely across the globe, so individuals vary in their predominant mood. Some enjoy a stable, sunny disposition, while others see the world through a darker lens. Such variation in the way we engage the world (psychiatrists call it temperament) is woven into the fabric of human behavior. Thus, we are talking here about the biology of the self in its deepest, most personal sense.

Psychiatric disorders are characterized by exaggerations of normal behavior, so if we experience a persistent and unusual change in mood ourselves, or observe it in someone else, we have cause for concern. Disorders of mood are pervasive, long-lasting emotional states. They are extreme emotions that color a person's outlook on life and affect behavior. For example, depression is an extreme form of melancholy or sadness accompanied by a lack of energy and a lack of emotion, while mania is an extreme form of elation and hyperactivity. In bipolar disorder, mood alternates between these two extremes.

In this chapter, we consider the role that emotion plays in our everyday life, in our sense of self. We then examine the characteristics of depression and bipolar disorder and what they tell us about ourselves. We explore several remarkable advances in brain science that point to the causes of depression and bipolar disorder and that have led to promising new treatments for these disorders. We examine the importance of psychotherapy for people with mood disorders, both separate from and in combination with drugs. Finally, we look at the contribution of genes to disorders of mood. These revelations underscore the vital link between studies of brain disorders and our understanding of how the healthy emotional brain works.

EMOTION, MOOD, AND THE SELF

Our emotions are coordinated by the amygdala, a structure that lies deep within each of the temporal lobes of the brain. The amygdala connects to several other structures in the brain, among them the hypothalamus and the prefrontal cortex. The hypothalamus regulates heart rate, blood pressure, sleep cycles, and other bodily functions involved in our emotional reactions. As such, it is the executor of emotion, including happiness, sadness, aggression, eroticism, and mating. The prefrontal cortex, the seat of executive function and self-esteem, regulates emotion and its influence on thought and memory. As we shall see, the connections among these structures account for the varied psychological and physical manifestations of mood disorders.

Emotion is part of the brain's early warning system and is intimately linked to the body's ancient survival mechanisms. As Charles Darwin first pointed out, emotions are part of a preverbal system of social communication that we share with other mammals. In fact, even with our extraordinary facility for language, we use emotion every day to communicate our desires to one another and to monitor our social environment. When our emotions signal that events are dangerous or not unfolding favorably, we experience feelings of anxiety, irritability, and vigilance, often followed by sadness. At the opposite end of the spectrum, falling in love and other positive emotions give us a wonderful feeling of renewed energy and optimism.

Our subjective emotional experience is constantly changing as our brain monitors the opportunities and stresses of a shifting social world and signals the appropriate coping response. Without these emotional assessments, we would experience the world as a series of random events with no point of reference—that is, with no sense of self.

Mood disorders are brain diseases that afflict the integrity of the self—that collection of vital emotions, memories, beliefs, and behaviors that shapes each of us as a unique human being. It is precisely because of emotion's central role in both our thinking and our feeling—and because we experience normal shifts in mood every day—that we have such difficulty identifying and accepting a mood disturbance as potentially abnormal. That same difficulty helps explain why people with mood disorders are frequently stigmatized. To put it simply, despite advances in science and medicine, many people are still inclined to view mood disorders as a personal weakness, as bad behavior, rather than as a set of illnesses.

DISORDERS OF MOOD AND THE ORIGINS OF MODERN PSYCHIATRY

Emil Kraepelin, whom we met in chapter 1, was a founder not only of modern scientific psychiatry, but also of psychopharmacology, the study of drugs' effects on mood, thought, and behavior. In 1883 he published *Compendium of Psychiatry*, the first edition of what would grow into his great multivolume *Textbook of Psychiatry*. In 1891 he began to teach at the University of Heidelberg and later moved to the University of Munich. Kraepelin held that mental illnesses are strictly biological and that they have a heritable basis. What's more, he insisted that psychiatric diagnoses be founded on the same criteria as diagnoses in other areas of medicine.

Kraepelin had set himself a difficult task. In his day, it was impossible to confirm diagnoses of psychiatric diseases at autopsy because such diseases do not leave dramatic marks on the brain, and brain-imaging technology wouldn't appear on the horizon until a century later. In the absence of biological markers and imaging, Kraepelin had to base his diagnoses on clinical observations of his patients.

To guide his observations, Kraepelin relied on the same three criteria used in general medicine: What are the symptoms of the disease? What is the course of the disease? What is the final outcome?

Applying those criteria to mental illnesses, Kraepelin distinguished two major groups of psychotic disorders: disorders of thought and disorders of mood. He called the disorders of thought *dementia praecox*—the dementia of young people—because they start earlier in life than other dementias, such as Alzheimer's, and he called the disorders of mood *manic-depressive illness* because they manifest themselves as either depressed or elevated feeling states. We now refer to dementia praecox as schizophrenia, and we refer to manic-depressive illness as bipolar disorder. We refer to depressed states alone, with no manic component, as major depression, or unipolar depression. The majority of people with depressive disorders are unipolar.

The distinctions that Kraepelin observed between the two major psychiatric disorders—schizophrenia and bipolar—have carried forward to this day. However, because recent genetic studies suggest that some genes may contribute to both types of disorders, we now realize that there may be overlap between them. There may also be overlap between these disorders and autism, which was recognized fully half a century after Kraepelin's classic work.

Disorders of thought and disorders of mood not only affect people differently, they run different courses and have different outcomes. Schizophrenia is characterized by cognitive decline that begins with the first episode of illness, usually in young adulthood, and continues throughout life, often with no remission. Mood disorders, in contrast, are most commonly episodic, with months to years elapsing between episodes. Major depression generally begins in the late teens and early twenties, whereas bipolar disorder usually begins in late adolescence. The average length of remission in major depression is about three months. This indicates that, at least initially, the changes in neural circuitry and brain function that lead to depression are reversible. As a person ages, the episodes of depression tend to last longer and the intervals of remission become shorter. People with a mood disorder can function very well during periods of remission, and the outcome of mood disorders is often more benign than that of schizophrenia.

Because they affect neural circuits in many regions of the brain, mood disorders also cause changes in energy, patterns of sleep, and thinking. Many depressed people, for example, have trouble getting to sleep and staying asleep; others sleep all the time, especially if they are more withdrawn than anxious. Sleep deprivation, which causes increased activity in the amygdala, can trigger manic episodes in some people with bipolar disorder.

Treatments for people with psychiatric disorders have improved by fits and starts since Philippe Pinel liberated the inmates of Salpêtrière hospital from their chains. A century elapsed before Pinel's insistence that psychiatric disorders are medical in nature and that heredity plays a role in them was carried forward by Kraepelin. It took equally long for Pinel's humane treatment of patients to reach fruition in psychotherapy. Since then, we have developed new forms of psychotherapy, new drug therapies, and a greater biological understanding of how these therapies act and interact. An essential component of treatment is understanding and accepting that psychiatric disorders are lifelong. As a result, people with mood disorders must be constantly aware of their feelings and their state of mind.

In this chapter we will examine depression and bipolar disorder separately, to see what mood disorders reveal about normal mood states.

DEPRESSION

Depression was first recognized in the fifth century B.C. by the Greek physician Hippocrates, who was one of the most influential physicians in history and is generally considered the father of Western medicine. Physicians at the time of Hippocrates did not believe that diseases affect particular organs of the body. Rather, they subscribed to the theory that all diseases are caused by an imbalance of the four "humors," or fluids, of the body: blood, phlegm, yellow bile, and black bile. Thus, Hippocrates thought that depression results from an excess of black bile in the body. In fact, the ancient Greek term for depression, *melancholia*, means "black bile."

The clinical features of depression were first, and perhaps best, summarized by William Shakespeare, that great observer of the human mind, whose Hamlet declares, "How weary, stale, flat, and unprofitable seem to me all the uses of this world." The most common symptoms of depression are feelings of persistent sadness and intense mental anguish, accompanied by feelings of hopelessness, helplessness, and worthlessness. Often, these feelings lead to withdrawal from the company of others; sometimes they lead to thoughts of, or attempts at, suicide. At any given time about 5 percent of the world's population suffers from major depression, including 20 million Americans. It is the primary cause of disability in people age fifteen through forty-five.

People with depression often describe feeling intense psychic suffering and isolation. In *Darkness Visible*, a memoir about his experience with depression, the American novelist and essayist William Styron wrote, "The pain is unrelenting, and what makes the condition intolerable is the foreknowledge that no remedy will come—not in a day, an hour, a month, or a minute."[1]

Today, we know that depression results not from black bile but from changes in brain chemistry. Still, we do not fully understand the mechanisms in the brain responsible for those changes. Scientists have made great strides, as we shall see, but depression is a complex disorder. In fact, depression is probably not one but several different disorders, with different degrees of severity and different biological mechanisms.

DEPRESSION AND STRESS

Stressful life events—the death of a loved one, loss of a job, a major move, or rejection in a love relationship—can trigger depression. At the same time, depression can cause or exacerbate stress. Andrew Solomon (fig. 3.1), a professor of clinical psychology at Columbia University and a superb writer, describes the onset of depression following several stressful events in his life:

> I had always thought of myself as fairly tough, fairly strong, and fairly able to cope with anything. And then I had a series of personal losses. My mother died. A relationship that I was in came to an end, and a variety of other things went awry. I managed to get through those crises more or less intact. Then, a couple of

Figure 3.1. Andrew Solomon

years later, I suddenly found myself feeling bored a lot of the time. . . . I remember particularly that, coming home and listening to the messages on my answering machine, I would feel tired instead of being pleased to hear from my friends, and I'd think, That's an awful lot of people to have to call back. I was publishing my first novel at the time, and it came out to rather nice reviews. I simply didn't care. All my life I had dreamed of publishing a novel, and now here it was, but all I felt was nullity. That went on for quite a while. . . .

Then . . . [e]verything began to seem like such an enormous, overwhelming effort. I would think to myself, Oh, I should have some lunch. And then I would think, But I have to get the food out. And put it on a plate. And cut it up. And chew it. And swallow it. . . . I knew that what I was experiencing was idiotic. It was nonetheless vivid and physical and acute, and I was helpless in its grip. As time went on, I found myself doing less, going outside

> less, interacting with other people less, thinking less, and feeling less.
>
> Then the anxiety set in. . . . The most acute hell of depression is the feeling that you will never emerge. If you can alleviate that feeling, the state, though miserable, is bearable. But, if someone were to say to me that I had to have acute anxiety for the next month, I would kill myself, because every second of it would be so intolerably awful. It is the constant feeling of being absolutely terrified and not knowing what it is that you're afraid of. It resembles the sensation you have if you slip or trip, the feeling you get when the ground is rushing up at you before you land. That feeling lasts about a second-and-a-half. The anxiety phase of my first depression lasted six months. It was incredibly paralyzing. . . .
>
> I got sicker and sicker until finally one day I woke up and actually thought that perhaps I'd had a stroke. I remember lying in bed and thinking that I'd never felt so bad in my life and that I should call someone. From my bed I looked at the telephone on my nightstand, but I could not reach out and dial a number. I lay there for four or five hours, just staring at the telephone. And finally it rang. I managed to answer it. I said, "I'm in terrible trouble." And that was when I finally sought antidepressants and began the serious treatment of my illness. . . .[2]

Depression and stress appear to set off the same biochemical changes in the body: they activate the neuroendocrine system's hypothalamic-pituitary-adrenal axis, prompting the adrenal gland to release cortisol, the body's primary stress hormone. While the release of cortisol for a short period is beneficial—it heightens our vigilance in response to a perceived threat—the long-term release of cortisol in major depression and chronic stress is harmful. It causes the changes in appetite, sleep, and energy that depressed and highly stressed people experience.

Excessive concentrations of cortisol destroy synaptic connections between neurons in the hippocampus, the region of the brain that is important in memory storage, and neurons in the prefrontal cortex, which regulates a person's will to live and influences a person's decision making and memory storage. The breakdown of synaptic connections in these regions leads to the flattening of emotion and to the impaired memory and concentration that accompany major depression and chronic stress. Many brain-imaging studies of people with depression have shown a decrease in the overall size and number of synapses between neurons in the prefrontal cortex and hippocampus; similar changes have been found in postmortem studies. Moreover, studies in mice and rats reveal that, when under stress, these animals also lose synaptic connections in the hippocampus and prefrontal cortex.

Animal models have given us valuable insights into the neural circuit of fear that underlies stress. Studies reveal that both instinctive fear and learned fear recruit the amygdala and the hypothalamus. The amygdala, as we know, determines what emotion is recruited at any given time, and the hypothalamus carries it out. When the amygdala calls for a fear response, the hypothalamus activates the sympathetic nervous system, which elevates heart rate, blood pressure, and secretion of stress hormones and regulates erotic, aggressive, defensive, and escape behavior.

These findings are all consistent with the idea that prolonged stress—which prompts the long-term release of cortisol and the consequent loss of synaptic connections—is an important component of depressive disorders, including the depressive phase of bipolar disorder.

THE NEURAL CIRCUIT OF DEPRESSION

Until quite recently psychiatric disorders were notoriously difficult to trace to particular regions in the brain. But today's brain-imaging technologies, specifically PET and functional MRI, have enabled scientists to identify at least some components of the neural circuit responsible for depression. By examining this circuit systematically in patients who have volunteered for studies, scientists have come to understand which patterns of neural activity are altered and can examine the effects of antidepressant drugs and psychotherapy on those abnormal patterns of activity. Moreover, recent brain-imaging technology has allowed scientists to identify biological markers in the brain that indicate which patients need just psychotherapy and which need both drug treatment and psychotherapy.

Helen Mayberg, a neurologist now at Emory University, found that the neural circuit of depression has several nodes, two of which are particularly critical: cortical area 25 (the subcallosal cingulate cortex), and the right anterior insula.[3] Area 25 is a region where thought, motor control, and drive come together. It is also rich in neurons that produce serotonin transporters—proteins that remove serotonin from the synapse. This is important because serotonin is a modulatory neurotransmitter released by a class of nerve cells to help regulate mood. Modulatory transmitters don't simply transmit an impulse from one cell to the next, but "tune" whole circuits or regions. Serotonin transporters are particularly active in depressed people and are partially responsible for lowering the concentration of serotonin in area 25. The second critical node, the right anterior insula, is a region where self-awareness and social experience come together. The anterior insula connects to the hypothalamus, which helps regulate sleep, appetite, and libido, and to the amygdala, the hippocampus, and the prefrontal cortex. The right anterior insula receives information from our senses about the physiological state of our body and, in response, generates emotions that inform our actions and decisions.

Another brain structure that is consistently implicated in both major depressive and bipolar disorders is the gyrus, or raised fold, of the anterior cingulate cortex. This structure runs parallel to the corpus callosum, the band of nerve fibers that connects the left and right hemispheres of the brain. The anterior cingulate gyrus is divided functionally into two regions. One region (the rostral and ventral subdivision) is thought to be involved in emotional processes and autonomic functions; it has extensive connections to the hippocampus, amygdala, orbital prefrontal cortex, anterior insula, and the nucleus accumbens, an important part of the brain's dopamine reward and pleasure circuit, as we will see in chapter 9. The other region (the caudal subdivision) is thought to be involved in cognitive processes and in the control of behavior; it connects with the dorsal areas of the prefrontal cortex, secondary motor cortex, and posterior cingulate cortex.

Both regions function abnormally in people with mood disorders, which accounts for their varied emotional, cognitive, and behavioral symptoms. The region concerned with emotion is consistently overactive during major depressive episodes and the depressive phase of bipolar disorder. Indeed, as we shall see, successful treatment with antidepressant drugs is correlated with decreased activity in a particular part of this region, the subgenual area of the anterior cingulate gyrus.

THE DISCONNECT BETWEEN THOUGHT AND EMOTION

At the same time that she found hyperactivity in area 25, Mayberg found underactivity in other parts of the prefrontal cortex of people with depression.[4] The prefrontal cortex, as we know, is responsible for concentration, decision making, judgment, and planning for the future. It connects directly to the amygdala, hypothalamus, hippocampus, and insular cortex, and each of these regions, in turn, connects directly to area 25. Conversations among these areas of our brain make use of emotion and thinking to help us plan our day and respond to the world around us in a healthy way.

Brain imaging has revealed several changes in the structure of the brain that may account for some of the symptoms that people with mood disorders experience. For example, imaging has shown that people with depression have an enlarged amygdala, and that people with depression, bipolar disorder, and anxiety disorders have increased activity in the amygdala. Scientists have suggested that increased activity of the amygdala may account for the hopelessness, sadness, and mental anguish that people with depression feel. Imaging has also found that, like many other disorders, depression may result in fewer and smaller synapses in the hippocampus. In fact, longer depressive episodes are correlated with reductions in the volume of the hippocampus. This correlation would account for the problems with memory that people with depression experience. Defective functioning of the hypothalamus, as revealed in imaging, may account in part for the loss of drive in people with depression, whether the drive for sex or the appetite for food. Finally, defective functioning of the insular cortex, a structure involved with bodily sensations, may account for why people with depression are without vitality, why they often feel dead inside.

Studies of depression suggest that whenever area 25 becomes hyperactive, the components of the neural circuit concerned with emotion are literally disconnected from the thinking brain, leading to a loss of personal identity. Mayberg's brain-imaging studies of depression reveal where these breaks in the circuit occur and help explain why depression can cause bodily sensations that patients can't place or consciously do anything about.[5]

TREATING PEOPLE WITH DEPRESSION

The most important reason to develop effective treatments for depression is to prevent suicide. Depression accounts for more than half of the forty-three thousand suicides that occur in the United States each year. Moreover, nearly 15 percent of people with depression commit suicide. That rate is much higher than the suicide rate among people with terminal illnesses, it equals the rate of homicide in the general U.S. population, and it has overtaken the rate of traffic fatalities in the United States. Although twice as many women suffer from depression as men, and women *attempt* suicide three times more often than men, men are three or four times as likely to actually kill themselves. The reason is that men tend to choose more aggressive methods—guns, jumping off bridges, throwing themselves under a subway train—and such methods are more likely to be fatal.

DRUG TREATMENTS

The first drugs used to treat people with depression were discovered by sheer accident. That accident not only proved providential for patients, it also provided the first insight into aspects of the biochemical disturbance underpinning depression.

In 1928 Mary Bernheim, a graduate student in the Department of Biochemistry at the University of Cambridge in England, discovered monoamine oxidase (MAO), an enzyme that breaks down a class of neurotransmitters known as monoamines.[6] (Neurotransmitters, as we have seen, are chemical messengers that neurons release into the synapses to communicate with other neurons.) Her discovery led to the introduction of a drug called iproniazid, which was used to treat people with tuberculosis. In 1951 doctors and nurses working on the tuberculosis ward of Sea View Hospital on Staten Island, New York, noticed that their patients taking iproniazid seemed less lethargic and much happier than those who were not taking the drug. Subsequent clinical trials revealed that iproniazid had antidepressant properties. Shortly thereafter, imipramine, a drug developed initially to treat people with schizophrenia, was also found to relieve symptoms of depression, by blocking the reuptake of monoamines into nerve terminals. *Reuptake* is a process that recycles the neurotransmitters and stops the signaling.

The antidepressant effects of iproniazid and imipramine suggested that monoamines were somehow involved in depression. But how?

Researchers discovered that monoamine oxidase breaks down and removes from the synapses two neurotransmitters: noradrenaline and serotonin. Without enough of these neurotransmitters, people experience symptoms of depression. The scientists reasoned that inhibiting the action of the enzyme that removes the monoaminergic transmitter from the synapse leaves more noradrenaline and serotonin in the synapses, thereby relieving the symptoms of depression. Thus, the idea of monoamine oxidase inhibitors as a treatment for depression was born. Later, researchers found that iproniazid and imipramine also lead to an increase in the size and number of synapses in the hippocampus and the prefrontal cortex, the brain regions in which synaptic connections are damaged by stress and depression.

Understanding how these two antidepressants work led to the development of the *monoamine hypothesis*, which holds that depression results from partial depletion of noradrenaline or serotonin, or both. This hypothesis also cleared up a mystery surrounding the drug reserpine, which had been used in the 1950s to treat high blood pressure and which induced depression in 15 percent of the people who took it. Reserpine, it turns out, also depletes noradrenaline and serotonin in the brain.

The monoamine hypothesis of depression was modified in the 1980s with the introduction of drugs such as fluoxetine (Prozac), which are known as *selective serotonin reuptake inhibitors* (SSRIs). These drugs increase concentrations of serotonin in the synapse by blocking the reuptake of serotonin; they do not act on noradrenaline. This finding led researchers to conclude that depression is related specifically to depletion of serotonin and not the depletion of noradrenaline.

In time, however, scientists realized that treating depression is more than a simple matter of flooding the synapses with serotonin. To begin with, boosting serotonin didn't help all patients get better. Conversely, reducing serotonin didn't consistently worsen symptoms in depressed people, nor did it produce depression in all healthy people. Moreover, antidepressant drugs such as Prozac increase serotonin very rapidly in depressed people, yet people don't show improvement in their mood or synaptic connections for weeks. While the monoamine hypothesis ultimately fell short of fully explaining the biology of depression, it spurred many good studies of the brain and helped clarify the important role that serotonin plays in the regulation of mood. In doing so, the hypothesis improved the lives of many people with depression.

Since selective serotonin reuptake inhibitors take about two weeks to take effect—a delay that can open the door to suicide attempts—and since a significant number of people do not respond at all to these reuptake inhibitors, new drugs were clearly needed. But despite intensive efforts, twenty years elapsed before a fast-acting drug emerged to treat people with depressive disorders.

That drug was ketamine, a veterinary anesthetic. Ketamine, whose mechanism of action was discovered by Ronald Duman and George Aghajanian at Yale,[7] acts within hours in people with treatment-resistant depression. What's more, the effect of that single dose can last for several days. Ketamine also appears to reduce suicidal thoughts and is now being explored as a possible short-term treatment for depressive episodes in people with bipolar disorder.

Ketamine works differently from traditional antidepressants. To begin with, it targets glutamate, not serotonin. To understand why this is important, we must first know that neurotransmitters fall into two categories: mediating and modulatory. *Mediating neurotransmitters* are released by a neuron at the synapse and act directly on the target cell, either exciting the target cell or inhibiting it. Glutamate is the most common excitatory transmitter, and GABA (gamma aminobutyric acid) is the most common inhibitory transmitter. *Modulatory neurotransmitters*, on the other hand, fine-tune the action of excitatory and inhibitory neurotransmitters. Dopamine and serotonin are modulatory neurotransmitters.

Because ketamine acts on the excitatory neurotransmitter glutamate, which directly affects the target cell, the drug reduces depression more quickly than drugs that act on the modulatory transmitter serotonin. In addition, ketamine prevents the transmission of glutamate from one neuron to the next by blocking a particular glutamate receptor on the target cell. Since a receptor blocked by ketamine can't bind glutamate, the neurotransmitter can't affect the target cell. The demonstration of the antidepressant effect of ketamine profoundly changed the way we think about depression.

Ketamine's beneficial effects reveal yet another mechanism contributing to depression. As we have seen, depression is caused not simply by insufficient serotonin and adrenaline but also by stress, which results in the release of excessive cortisol, damaging neurons in the hippocampus and the prefrontal cortex. As it happens, high concentrations of cortisol also cause an increase in glutamate, and large doses of glutamate damage neurons in exactly the same areas of the brain.

Almost all antidepressants, including ketamine, promote the growth of synapses in the hippocampus and prefrontal cortex, thus countering the damage caused by cortisol and glutamate and providing an additional explanation for why these drugs are so effective. Moreover, in rodents ketamine acts quickly to induce the growth of synapses and to reverse the atrophy caused by chronic stress. As a result, the discovery of ketamine has been hailed as the most important advance in depression research in the last half century. However, because it produces side effects such as nausea, vomiting, and disorientation, ketamine cannot be taken over the long term and therefore cannot replace the selective serotonin reuptake inhibitors. Instead, because of its rapid action, ketamine is used to lessen the risk of suicide during the approximately two weeks required for the serotonin-enhancing drugs to take effect.

PSYCHOTHERAPY: THE TALKING CURE

Psychotherapy is an integral part of treatment for most people with psychiatric illnesses. Put simply, it is a verbal exchange between a patient and a therapist within a supportive relationship. While various forms of psychotherapy may have somewhat different theoretical bases, they all share this essential element. Psychotherapy has been used to treat patients for over a century, but scientists are only now beginning to understand how it works on the brain.

The first form of psychotherapy was psychoanalysis, which originated with Josef Breuer, a senior colleague of Freud's at the Vienna School of Medicine. In 1895 Freud joined Breuer in publishing a paper about a patient, Anna O., who was suffering from paralysis on the left side of her body—a paralysis that had no neurological basis.[8] Breuer encouraged Anna O. to talk at random about her memories, fantasies, and dreams. In the course of this *free association*, as he later called it, she remembered traumatic events. The recovery of those memories led to relief from her paralysis.

Freud was much impressed with this case. He picked up Breuer's technique and used it to obtain insights into his own patients. From their fantasies and memories Freud inferred that the origins of mental illness lie in infancy and early childhood. Three modern scholars of psychoanalysis, Steven Roose of Columbia University College of Physicians and Surgeons, Arnold Cooper of Weill Cornell Medical Center, and Peter Fonagy of University College London, point to three key observations of Freud's that are central to psychoanalysis.[9]

First, children have sexual and aggressive behavioral instincts. The social prohibitions that hold these instinctual needs in check begin early in life and carry onward into adulthood. In other words, sexuality and aggression do not arise in adulthood; they are present in infancy.

Second, children suppress and render unconscious the conflicts between early needs and prohibitions, as well as early traumas. These repressed feelings may result in symptoms of mental illness in adulthood. In the course of free association during psychoanalysis, the patient liberates his or her repressed conflicts. The therapist's interpretations of those revelations can help resolve the conflicts, thereby relieving the patient's mental symptoms.

Third, the patient's relationship with the therapist reenacts the patient's early relationships. This reenactment is called *transference*. Transference and the interpretation of the transference by the therapist play a central role in the therapeutic process.

Psychoanalysis heralded a new method of psychological investigation, a method based on free association and interpretation. Freud taught analysts to listen carefully to patients in ways that no one had before. He also outlined a provisional way of making sense out of patients' seemingly unrelated and incoherent associations.

While psychoanalysis has historically been scientific in its aims, it has rarely been scientific in its methods (see chapter 11). Indeed, Freud and the original founders of psychoanalysis made few serious attempts to establish proof of the efficacy of psychotherapy. That way of thinking changed in the 1970s, when Aaron Beck, a psychoanalyst at the University of Pennsylvania, set out to test Freud's ideas about depression.

Freud held that depressed people feel hostile toward someone they love, yet they have difficulty harboring negative feelings about a person who is important to them. They therefore repress the negative feelings and unconsciously direct them inward. This anger ultimately leads to feelings of worthlessness and low self-esteem, which are characteristic of depression.

Yet Beck found that his depressed patients actually exhibited less hostility than his other patients. Instead, the depressed patients consistently saw themselves as losers, had unrealistically high expectations of themselves, and handled even the simplest disappointment badly. This pattern of thinking reflects a disorder of cognitive style, of how we perceive ourselves in the world.

Beck wondered if identifying those negative beliefs and thought processes and then helping the patient replace them with more positive thoughts might relieve depression without having to deal with specific unconscious conflicts. He tested his idea by presenting patients with evidence of their accomplishments, achievements, and successes, thus challenging their negative views of themselves. His patients often improved with remarkable speed, feeling and functioning better after just a few sessions.

This positive result encouraged Beck to develop a short, systematic psychological treatment for depression based on a patient's cognitive style and distorted way of thinking. He called the treatment *cognitive behavioral therapy*. Once he was certain that it worked repeatedly, he wrote a manual on it so that other people could carry out the same treatment.[10] Finally, he conducted outcome studies.

The outcome studies showed that for mild and moderate depression, cognitive behavioral therapy was better than a placebo and as good as, if not better than, antidepressant drugs. For severe depression, the therapy was not as good as an antidepressant; however, therapy and the antidepressant were synergistic—that is, the two treatments together benefited the patient more than either treatment by itself.[11]

Cognitive behavioral therapy has had a mighty impact on psychiatry and on psychoanalytical thought. It showed that a complex process like psychotherapy can be studied and that its outcomes can be evaluated. As a result, psychotherapy is now being tested empirically.

Psychiatrists used to think that psychotherapy and drugs work in different ways, that psychotherapy acts on our mind and drugs act on our brain. Now they know better. The interaction between therapist and patient can actually change the biology of the brain. This finding should not be surprising. My own work has shown that learning leads to anatomical changes in the connections between neurons. This anatomical change underlies memory—and psychotherapy, after all, is a learning process.

Thus, insofar as psychotherapy produces persistent changes in behavior, it is also producing changes in the brain. In fact, studies are now giving us a better idea about what kinds of psychotherapy work best and for what kinds of patients.

COMBINING DRUGS WITH PSYCHOTHERAPY

All pharmacological treatments come with undesired side effects, ranging from annoying to life-threatening, and as a result patients often discontinue use of these drugs. Psychotherapy, which is known to be effective, does not have such side effects. Thus, for many people with depression, the best treatment is a combination of drugs and psychotherapy.

In the 1990s clinical investigators such as Beck figured out how to use drugs and psychotherapy synergistically. While medication helps to restore the balance of chemicals in the brain, psychotherapy provides a consistent, supportive, and healthy relationship with a therapist. These are the key ingredients to turning mental illness around and enabling people to live a fulfilling, productive life.

Kay Redfield Jamison, who is co-director of the Mood Disorders Center at The Johns Hopkins School of Medicine and who herself has bipolar disorder, strongly agrees with this point. In her book *An Unquiet Mind*, she writes that psychotherapy "makes some sense of the confusion, reins in the terrifying thoughts and feelings, returns some control and hope and possibility of learning from it all. Pills cannot, do not, ease one back into reality."[12]

Andrew Solomon concurs:

> Once I had begun to return to some reasonable facsimile of myself . . . I had to figure out what triggered my episodes and how to control them. This I did with the analytically trained therapist with whom I had begun working. . . . Once you have been depressed, and particularly once you have allowed medication to reshape your mental states, you need to understand who you are at the most fundamental level. . . .
>
> I now have a psychopharmacologist and a psychoanalyst, and I would not be who I am today without their work and without the work I have done with them both. The fashion for biological explanations of depression seems to miss the fact that chemistry has a different vocabulary for a set of phenomena that can also be described psychodynamically. Neither our pharmacology nor our analytic insight is advanced enough to do all the work; to approach the problem of depression from both angles is to figure out not only how to recover, but also how to live the life that must follow on recovery.[13]

In a recent study of people with depression, Mayberg provided each person with either cognitive behavioral therapy or an antidepressant medication. She found that people who started with below-average baseline activity in the right anterior insula responded well to cognitive behavioral therapy but not to the antidepressant. People with above-average activity responded to the antidepressant, but not to cognitive behavioral therapy. Thus, Mayberg found that she could predict a depressed person's response to specific treatments from the baseline activity in the right anterior insula.[14]

These results show us four very important things about the biology of brain disorders. First, the neural circuits disturbed by psychiatric disorders are complex. Second, we can identify specific, measurable markers of a brain disorder, and those biomarkers can predict the outcome of two different treatments: psychotherapy and medication. Third, psychotherapy is a biological treatment; it produces detectable, lasting physical changes in our brain. And fourth, the effects of psychotherapy can be studied empirically.

Many psychotherapists have been slow to investigate the empirical basis of their treatment, in part because a number of them believe that human behavior is too difficult to study in scientific terms. The finding by Mayberg that cognitive behavioral therapy is a biological treatment now provides an opportunity for evaluating the outcome of psychotherapy in a rigorously objective manner.

BRAIN STIMULATION THERAPIES

Some people with depression do not respond to drugs or psychotherapy. For many of these people, therapies such as electroconvulsive therapy and deep-brain stimulation have proved beneficial.

Electroconvulsive (shock) therapy gained a bad reputation during the 1940s and '50s because patients were given high doses of electricity without any anesthesia, resulting in pain, fractured bones, and other serious side effects. Today, electroconvulsive therapy is painless. It is administered after the patient has been given general anesthesia and a muscle relaxant, it uses small electric currents to induce a brief seizure, and it is often very effective. Many patients have six to twelve sessions over a period of several weeks. Scientists are still not clear about how it works, but it is thought to relieve depression by producing changes in the chemistry of the brain. Unfortunately, the effects of electroconvulsive therapy usually do not last very long.

In the 1990s deep-brain stimulation was refined to treat people with Parkinson's disease by Mahlon DeLong at Emory University and Alim-Louis Benabid at the Joseph Fourier University in Grenoble, France. In this treatment, surgeons place an electrode in the dysfunctional region of a neural circuit and implant a device elsewhere in the patient's body that sends high-frequency electrical impulses into the region—much as a pacemaker regulates the heartbeat. The impulses block the firing of neurons whose abnormal signals cause the symptoms of Parkinson's disease.

Mayberg was familiar with these advances and thought that slowing the firing rate of neurons in area 25 might relieve the symptoms of depression. She used deep-brain stimulation in the anterior insula region to treat twenty-five people whose depression was resistant to treatment. She collaborated with a team of neurosurgeons, first at the University of Toronto and then at Emory, who implanted the electrodes. When she turned on the electricity in the operating room, she saw almost immediate changes in the patients' mood. The patients no longer felt the unending psychic pain characteristic of depression. Moreover, the other symptoms of depression gradually lifted as well. People recovered and were stabilized for the long term.[15]

BIPOLAR DISORDER

Bipolar disorder is characterized by extreme changes in mood, thought, energy, and behavior that generally alternate between depression and mania. These alternating moods distinguish bipolar disorder from major depression.

Manic episodes are characterized by an elevated, expansive, or irritable mood, together with several other symptoms, including heightened activity, racing thoughts, impulsiveness, and decreased need for sleep. These episodes are often associated with high-risk behaviors such as substance abuse, sexual promiscuity, excessive spending, or even violence. During a manic episode, people may say and do things that strain their relationships with others. They may get into trouble with the law or at work. Manic episodes can be frightening, both for people with bipolar disorder and for the people close to them.

About 25 percent of people with major depression go on to experience a manic episode. The initial manic episode is usually triggered by a personal situation, an environmental circumstance, or both. Common triggers include stressful life events, whether positive or negative; conflict or stressful relations with other people; disrupted routine or patterns of sleep; overstimulation; and medical illness. The manic episode is then followed by a depressive episode. While bouts of depression typically recur in any form of depression, they recur twice as often in bipolar disorder. And since bipolar disorder consists of alternating periods of mania and depression, this means the manic episodes recur equally often.

Once the first manic episode is initiated—usually at the age of seventeen or eighteen—the brain is changed in ways we do not yet understand, such that even minor events can trigger a later manic episode. After the third or fourth manic episode, a trigger may not be required. As a person with bipolar disorder grows older, the disease advances and the intervals between episodes may become shorter, particularly if he or she discontinues treatment.

Bipolar disorder affects about 1 percent of Americans, or more than 3 million people. While depression affects more women than men, bipolar disorder affects men and women equally. The disorder takes several forms, but the most common forms are known as bipolar I and bipolar II. People with bipolar I disorder have manic episodes, and sometimes cross over into psychosis with symptoms such as delusions and hallucinations, whereas people with bipolar II disorder have less severe, *hypomanic* episodes. Some people experience symptoms of both mania and depression at the same time, a condition known as a mixed state.

We do not know exactly what causes bipolar disorder, but we do know that its origins are complex and involve genetic, biochemical, and environmental factors. We are all subject to fluctuating moods: an exciting event may cause us to feel euphoric, while an unpleasant one may make us feel cast down. Most of us return to the normal state in a short time. Yet the same event may cause a person with bipolar disorder to plunge into extreme depression or mania for a long time. Two risk factors are particularly important in bipolar disorder: first, a genetic predisposition, as indicated by a sibling or a parent with the disorder; and second, periods of great stress.

The depressive episodes in bipolar disorder are similar to those in major depression. Thus the research carried out on the biology of major depression—the critical role of stress, the neural circuit of depression, the disconnect between thought and emotion, the action of antidepressants, and the importance of psychotherapy—applies to the depressive phase of bipolar disorder as well. Unfortunately, our understanding of the molecular underpinnings of mania is not as advanced as our understanding of the underpinnings of depression.

TREATING PEOPLE WITH BIPOLAR DISORDER

People with bipolar disorder may not see the need for continued treatment, particularly during a manic phase. It's very difficult, for example, to convince an eighteen-year-old who's staying up all night—full of energy, full of seemingly great ideas, thinking fast and furiously—that he or she is sick. But as the mania progresses, the person can become disorganized, psychotic, and self-destructive.

Kay Jamison (fig. 3.2), whom we met earlier, first realized that she was ill when she was about seventeen years old and a senior in high school. She has described her bipolar illness and the interaction of medication and psychotherapy in treating it:

Figure 3.2. Kay Redfield Jamison

> There is a particular kind of pain, elation, loneliness, and terror involved in this kind of madness. When you're high it's tremendous. The ideas and feelings are fast and frequent like shooting stars, and you follow them until you find better and brighter ones. Shyness goes, the right words and gestures are suddenly there, the power to captivate others a felt certainty. There are interests found in uninteresting people. Sensuality is pervasive and the desire to seduce and be seduced irresistible. Feelings of ease, intensity, power, well-being, financial omnipotence, and euphoria pervade one's marrow. But, somewhere, this changes. The fast ideas are

> far too fast, and there are far too many; overwhelming confusion replaces clarity. Memory goes. Humor and absorption on friends' faces are replaced by fear and concern. Everything previously moving with the grain is now against—you are irritable, angry, frightened, uncontrollable, and enmeshed totally in the blackest caves of the mind. You never knew those caves were there. It will never end, for madness carves its own reality.[16]

Imaging studies of brain function have shown widespread differences between healthy brains and the brains of people with bipolar disorder. That's no surprise. But if manic episodes are what distinguishes bipolar disorder from depression, then we should see additional or different changes in the brains of people with bipolar disorder, changes that cause the symptoms of mania and the cycling from one state to the other. In fact, however, convincing differences have been difficult to document. The best insights have come from attempts to understand how lithium, the most successful treatment for manic illness, affects the brain.

In the second century B.C. the Greek physician Soranus treated his manic patients with alkaline waters now known to be high in lithium. The benefit of lithium was rediscovered in 1948 by the Australian psychiatrist John Cade, who noticed that the substance made guinea pigs temporarily lethargic. Cade formally introduced lithium into the modern treatment of bipolar disorder in 1949, and it has been used ever since.

Unlike other medications used to treat psychiatric illnesses, lithium is a salt; consequently, it does not bind to a receptor on the surface of a neuron. Rather, it is actively transported into the neuron through sodium ion channels in the cell membrane that open in response to an external stimulus (see chapter 1). When a sodium ion channel opens, both sodium and lithium move into the cell. The sodium is subsequently pumped out, but the lithium remains inside. There, lithium may stabilize mood swings by affecting the action of neurotransmitters, either directly or through interaction with a second messenger system.

As we have seen, neurotransmitters bind to receptors on the cell membrane. This activates second messenger systems, which transmit signals from the receptors to molecules inside the neuron. Lithium may blunt the activation of second messenger systems, thus reducing signal transmission. Lithium may also damp down a neuron's responsiveness to neurotransmitters inside the cell. This could explain why lithium works so effectively in bipolar disorder: it may decrease a neuron's sensitivity to both external and internal stimuli. In addition, lithium affects the modulatory neurotransmitters serotonin and dopamine as well as the mediating neurotransmitter GABA. Thus, its efficacy may be attributable to its wide-ranging neurobiological effects rather than to a single mechanism.

Another possible way that lithium exerts its beneficial effects is by resetting ionic homeostasis in overly active neurons. The idea here is that lithium returns neurons to their resting state by increasing or decreasing their sensitivity to stimuli. Again, lithium may act directly on the surface receptors of neurons or through its interaction with intracellular second messenger systems.

One fascinating aspect of lithium treatment for mania is that it does not take effect for several days, and its effects do not disappear immediately after treatment is discontinued.

Today, bipolar disorder is treated with a combination of mood-stabilizing drugs and psychotherapy. Psychotherapy helps people with bipolar disorder recognize the particular emotional and physical situations that trigger depressive or manic episodes and emphasizes the importance of managing and reducing stress. Depressive episodes of bipolar disorder that are not contained by mood stabilizers such as lithium, atypical antipsychotics, or antiepileptic drugs are treated with antidepressants. While lithium reduces the severity and frequency of manic episodes in many patients, not everyone with bipolar disorder responds to it. Moreover, lithium has unpleasant side effects. We therefore need to find even better treatments.

MOOD DISORDERS AND CREATIVITY

The association between mood disorders and creativity, particularly the relationship between creativity and bipolar disorder, has been noted throughout history, from ancient Greece to the modern era. Vincent van Gogh, for example, suffered from depression during much of his adult life and committed suicide at the age of thirty-seven. Yet despite suffering from severe episodes of psychotic depression and mania during the last two years of his life, he produced three hundred of his most important works during that time. These works have proven to be important in the history of modern art because van Gogh used color not to convey the reality of nature but arbitrarily, to convey mood.

Empirical studies of contemporary artists and writers have found high rates of bipolar disorder among these groups. We will further consider the relationship between creativity and mood disorders in chapter 6.

THE GENETICS OF MOOD DISORDERS

For the most part, our genes determine whether we are likely to develop a mood disorder. As we saw in chapter 1, studies of identical twins who are reared apart—the best way to separate nature from nurture—indicate that if one twin has bipolar disorder, the other twin has a 70 percent chance of developing it. For major depression, the likelihood is 50 percent.

Scientists have recently discovered that complex brain disorders such as depression, bipolar disorder, schizophrenia, and autism share some genetic variants that increase the risk of developing one of these disorders. Thus, bipolar disorder emerges from an interaction of genetic and developmental factors with environmental factors. Scientists have also found two genes that may create a risk for both schizophrenia and mood disorders. Thus, it is clear that no single gene significantly affects the development of either bipolar disorder or schizophrenia. Many different genes are involved, and they work together with environmental factors in a complicated way. We discuss these and other findings of genetic research in more detail in chapter 4.

Recently, an international team analyzed genetic information from 2,266 people with bipolar disorder and 5,028 comparable individuals without the disorder. They merged their information with information on thousands of other individuals from previous studies. Altogether, the database included genetic material from 9,747 people with bipolar disorder and 14,278 people without the disorder.

The researchers analyzed about 2.3 million different regions of DNA. Their search led them to five regions that appeared to be connected to bipolar disorder.[17] Two of the regions contain new candidate genes that are likely to predispose a person to bipolar disorder, one on chromosome 5 and one on chromosome 6; the remaining three regions, previously suspected of a connection, were confirmed to be linked to the disorder. One of the newly discovered genes, *ADCY2*, was of particular interest. This gene oversees the production of an enzyme that facilitates neural signaling, a finding that fits very well with the observation that information transfer in certain regions of the brain is impaired in people with bipolar disorder.

Identifying the genes that make us susceptible to bipolar disorder, as this team did, is an important step in understanding how mood disorders develop. Once we understand their biological foundations, we can begin work on more effective and accurately targeted treatments. We can also recognize individuals at risk, leading to earlier intervention and an understanding of the environmental factors that interact with genes to create mood disorders. Finally, by understanding the biology of mood disorders we also begin to understand the biological underpinnings of the normal mood states that underlie our everyday emotional well-being.

LOOKING AHEAD

Our understanding of the genetics of depression and bipolar disorder is still in the early stages. These are, after all, very complex diseases. They disrupt the connections between the brain structures responsible for emotion, thought, and memory—connections that are crucial to our sense of self. This is why people with mood disorders experience such an array of psychological and physical symptoms. Only recently have neuroscientists been able to see, in real time, what goes on in the brains of people with these disorders, thus offering the possibility of correlating genetics, brain physiology, and behavior.

Nevertheless, tremendous advances have been made in other areas of research, particularly research on depression—finding the neural circuit for depression, using deep-brain stimulation to change the firing of neurons in that circuit, viewing the disconnect between the brain structures responsible for emotion and for thought, and understanding the biological nature of psychotherapy. These and other advances have led to improved treatments for people with mood disorders.

Today, with constant vigilance, proper treatment, and expert, compassionate assistance from informed clinicians, most people with mood disorders can regain and maintain emotional equilibrium and hold their lives together. With understanding on the part of family members and friends—understanding of both the patient's experience and the science of the illness—damage to relationships can be avoided or repaired. As a result of our gaining a biological understanding of the self, mood disorders have become treatable illnesses.

4

THE ABILITY TO THINK AND TO MAKE AND CARRY OUT DECISIONS: SCHIZOPHRENIA

Schizophrenia probably begins before birth, but it usually doesn't become apparent until late adolescence or early adulthood. When it does appear, schizophrenia often has devastating effects on thinking, volition, behavior, memory, and social interaction—the underpinnings of our sense of self—just at the time in their lives when young people are becoming independent. Like depression and bipolar disorder, schizophrenia is a complex psychiatric disorder affecting numerous regions of the brain and ultimately undermining the integrity of the self.

The biology of schizophrenia is particularly difficult to sort out because of the disorder's wide-ranging effects on the brain and behavior. This chapter presents what brain scientists have been able to discover about schizophrenia thus far: what circuits it disrupts in the brain, what treatments are available to patients, and what genetic and developmental components underlie the disorder. The emerging view of schizophrenia as a neurodevelopmental disorder that, unlike autism, manifests itself later in life has arisen from the considerable genetic research done on the disease.

Recent technical advances in genetics and brain imaging have given scientists new insights into the biology of schizophrenia. Based on those advances, we are now beginning to understand how schizophrenia affects the brain and to develop animal models that allow us to test specific hypotheses and to explore how the disease begins. These recent advances may provide a path to early intervention and treatment.

THE CORE SYMPTOMS OF SCHIZOPHRENIA

Schizophrenia produces three types of symptoms, each resulting from disturbances in a different region of the brain. This makes schizophrenia a particularly difficult disorder to understand and to treat.

The positive symptoms of schizophrenia—called "positive" not because they are good but because they represent new types of behavior for the person who has them—are the symptoms most frequently associated with the disease and the ones patients often recognize first. Positive symptoms reflect disordered volition and thinking. Disordered thought detaches a person from reality, leading to altered perceptions and behavior, such as hallucinations and delusions. These psychotic symptoms can be terrifying, not just for people who experience them but also for people who witness them. They are also a major cause of the stigma attached to people with schizophrenia.

The English artist Louis Wain conveyed his experience of the positive symptoms of schizophrenia (notably, altered perception) in his drawings of cats (fig. 4.1). As Kraepelin appreciated, and as we shall see in chapter 6, remarkable artistic capabilities sometimes manifest themselves for the first time in people who have developed schizophrenia. Thus, artists who become schizophrenic may continue to paint, and some people with schizophrenia who never painted before may take up painting as a means of giving voice to their feelings.

Hallucinations, the most common positive symptom, can be visual or auditory. Auditory hallucinations are very troubling: patients hear voices saying harshly critical, sometimes abusive things to them. The voices may cause them to harm themselves or others. Delusions, or false beliefs with no basis in fact, are also common. Of the several categories of delusions, the most common type is paranoid delusions. Patients often feel as though other people are out to get them, or following them, or trying to harm them. It is not uncommon for patients to believe that someone is trying to poison them, particularly with their medications.

Another very common type of delusion involves reference, or control.

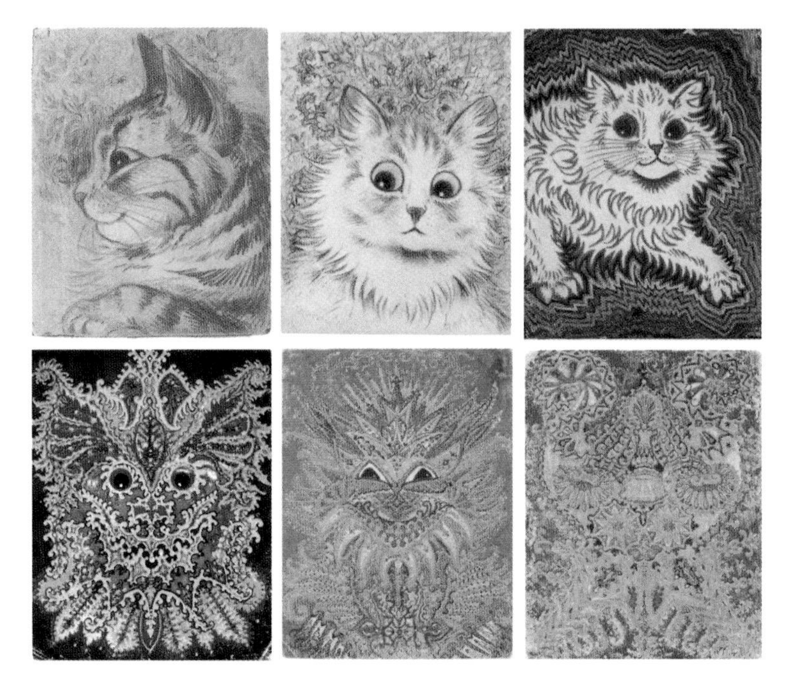

Figure 4.1. Drawings of cats by the artist Louis Wain (1860–1939), who had schizophrenia

Patients feel that they're receiving special messages, just for them, from the television or the radio; they often feel that other people can control their minds. Finally, patients may have delusions of grandeur, the feeling of having special powers.

The negative symptoms of schizophrenia—social withdrawal and lack of motivation—are typically present before the positive symptoms, but they are generally overlooked until a person experiences a psychotic episode. Social withdrawal may not entail actually avoiding people but rather being walled off and wrapped up in a separate world. Lack of motivation is evident in listlessness and apathy.

The cognitive symptoms of schizophrenia reflect problems with volition, with the executive functions involved in organizing one's life, and with working memory (a form of short-term memory), as well as early features of dementia. Patients are sometimes unable to gather their thoughts or to follow a train of thought. In addition, they may be unable to do the everyday things needed to be successful at work or to sustain relationships with others. As a result, they have great difficulty holding a job or marrying and raising children.

Brain scans of untreated people with schizophrenia reveal, over time, a subtle, but perceptible, loss of gray matter, which contains the cell body and dendrites of neurons in the cerebral cortex. This loss of gray matter, which contributes to the cognitive symptoms of schizophrenia, is thought to result from excessive pruning of dendrites during development, which leads to loss of synaptic connections among neurons, as we shall see later in this chapter.

To get a sense of how completely these symptoms of schizophrenia can loosen our hold on reality and sabotage our independence and sense of self, let us turn to someone who has the disorder: Elyn Saks (fig. 4.2), a professor of law at the University of Southern California and founder of the Saks Institute for Mental Health Law, Policy, and Ethics. In 2007 Saks published a book titled *The Center Cannot Hold*, in which she presents a frank and moving portrait of her experience of schizophrenia as well as a plea that we not impose limitations on people with schizophrenia but rather allow them to find their own limits. In September 2015

Figure 4.2. Elyn Saks

she was awarded a MacArthur Foundation "genius" grant. She described her terrifying initial psychotic experience:

> It's ten o'clock on a Friday night. I am sitting with my two classmates in the Yale Law School Library. They aren't too happy about being here; it's the weekend, after all—there are plenty of other fun things they could be doing. But I am determined that we hold our small-group meeting. We have a memo assignment; we have to do it, have to finish it, have to produce it, have to . . . Wait a minute. No, wait. "Memos are visitations," I announce. "They make certain points. The point is on your head. Have you ever killed anyone?"
>
> My study partners look at me as if they—or I—have been splashed with ice water. "This is a joke, right?" asks one. "What are you talking about, Elyn?" asks the other.
>
> "Oh, the usual. Heaven, and hell. Who's what, what's who. Hey!" I say, leaping out of my chair. "Let's go out on the roof!"
>
> I practically sprint to the nearest large window, climb through it, and step out onto the roof, followed a few moments later by my reluctant partners in crime. "This is the real me!" I announce, my arms waving above my head. "Come to the Florida lemon tree! Come to the Florida sunshine bush! Where they make lemons. Where there are demons. Hey, what's the matter with you guys?"
>
> "You're frightening me," one blurts out. A few uncertain moments later, "I'm going back inside," says the other. They look scared. Have they seen a ghost or something? And hey, wait a minute—they're scrambling back through the window.
>
> "Why are you going back in?" I ask. But they're already inside, and I'm alone. A few minutes later, somewhat reluctantly, I climb back through the window, too.
>
> Once we're all seated around the table again, I carefully stack my textbooks into a small tower, then rearrange my note pages. Then I rearrange them again. I can see the problem, but I can't see its solution. This is very worrisome. "I don't know if you're having the same experience of words jumping around the pages as I am," I say. "I think someone's infiltrated my copies of the cases. We've got to case the joint. I don't believe in joints. But they

do hold your body together." I glance up from my papers to see my two colleagues staring at me. "I . . . I have to go," says one. "Me, too," says the other. They seem nervous as they hurriedly pack up their stuff and leave, with a vague promise about catching up with me later and working on the memo then.

I hide in the stacks until well after midnight, sitting on the floor muttering to myself. It grows quiet. The lights are being turned off. Frightened of being locked in, I finally scurry out, ducking through the shadowy library so as not to be seen by any security people. It's dark outside. I don't like the way it feels to walk back to my dorm. And once there, I can't sleep anyway. My head is too full of noise. Too full of lemons, and law memos, and mass murders that I will be responsible for. I have to work. I cannot work. I cannot think.[1]

HISTORY OF SCHIZOPHRENIA

As we learned in chapter 3, Emil Kraepelin, the founder of modern scientific psychiatry, divided the major psychiatric illnesses into disorders of mood and disorders of thought. He was able to make this distinction because he brought to his studies of mental illness not only very astute clinical observations but also his training in the laboratory of Wilhelm Wundt, the pioneer of experimental psychology. Throughout his career Kraepelin strove to base the concepts of psychiatry on sound psychological research.

Kraepelin called the primary disorder of thought dementia praecox, the dementia of young people, because it starts earlier in life than Alzheimer's dementia. Almost immediately, the Swiss psychiatrist Eugen Bleuler took issue with the term. Bleuler thought dementia was only one component of the disease. Moreover, some of his patients had developed the disease later in life. Others functioned well after many years with the disease: they were able to work and have a family life. For these reasons, Bleuler called the disease *the schizophrenias*. He saw schizophrenia as a splitting of the mind—a disorientation of feelings from cognition and motivation—and he used the plural noun to recognize the several disorders embedded in this category. Bleuler's ideas are fundamental to our understanding of the disease, and his definition still holds.

TREATING PEOPLE WITH SCHIZOPHRENIA

Schizophrenia is not a rare disorder. It affects about 1 percent of people worldwide and roughly 3 million people in the United States. It strikes without regard to class, race, gender, or culture, and it varies greatly in severity. Many people with severe schizophrenia have difficulty forming or sustaining personal relationships, working, or even living independently. On the other hand, some people with milder forms of the disorder, such as the writer Jack Kerouac, the Nobel Prize winner in Economics John Nash, and the musician Brian Wilson, have had notable careers. Their symptoms are mostly kept in check by treatment with drugs and psychotherapy.

The drugs developed to treat people with schizophrenia initially focused on alleviating the positive symptoms of the disorder—that is, the psychotic symptoms: hallucinations and delusions. Antipsychotic drugs have been quite effective; in fact, most of the drugs we have today will alleviate positive symptoms to some extent for up to 80 percent of people with schizophrenia. However, antipsychotics are not very effective against the negative and cognitive symptoms of the disorder—and those symptoms can be the most pernicious and debilitating for patients.

Psychotherapy is also an essential treatment for people with schizophrenia. Interestingly, psychotherapy is now being used preemptively as well, for both the cognitive and the negative symptoms, to try to prevent the onset of psychotic symptoms in adolescents and young adults identified as being at risk. One of the many things that psychotherapy can accomplish is to help patients realize that they have a disorder, a disease: they are not a bad person but a good person suffering from delusions or hallucinations.

BIOLOGICAL TREATMENTS

Scientists got an initial glimpse into the biology of schizophrenia in the same way they got their initial look at the biology of depression—when the first effective drug appeared. In each case, that first drug emerged by chance, from drugs designed to work on another problem.

Paul Charpentier, a French chemist working for the pharmaceutical firm Rhône-Poulenc, had begun work on an antihistamine that he hoped would be effective against allergies but without producing the numerous side effects of existing antihistamines. The drug he developed in 1950 was called Thorazine (its generic name is chlorpromazine). As Thorazine went into clinical trials, everyone was amazed at its effect: it made people calmer, much more relaxed.

Seeing Thorazine's calming effects, Pierre Deniker and Jean Delay, two French psychiatrists, decided to give the drug to their psychotic patients. It was a magic bullet, particularly for their patients with schizophrenia. By 1954, when the U.S. Food and Drug Administration approved the drug, 2 million people in the United States alone had been treated with Thorazine. A great many of them were able to leave state mental hospitals.

Thorazine was originally thought to act as a tranquilizer, calming patients without sedating them unduly. However, by 1964 it became clear that Thorazine and related drugs produce specific effects on the positive symptoms of schizophrenia: they mitigate or abolish delusions, hallucinations, and some types of disordered thinking. Moreover, if patients take them during periods of remission, these antipsychotic drugs tend to reduce the rate of relapse. Yet the drugs have significant side effects, including neurological symptoms characteristic of Parkinson's disease. People taking the drugs develop a tremor of their hands, bend forward when they walk, and experience rigidity in their body.

Scientists eventually developed new drugs with fewer and less severe neurological side effects. Those drugs include clozapine, risperidone, and olanzapine, and they all are effective at controlling positive symptoms of the disease. Only clozapine is considered to be more effective than the earlier antipsychotics in treating the negative symptoms and cognitive defects of schizophrenia, and then only marginally so. The newer drugs are referred to as "atypical" antipsychotics because they all produce fewer Parkinson's-like side effects than the earlier, "typical" drugs.

The first clue to how typical antipsychotics work came from analysis of their neurological side effects. Since these drugs produce the same effects on movement as Parkinson's disease, which is caused by a deficiency in the modulatory neurotransmitter dopamine, scientists reasoned that the drugs might act by reducing dopamine in the brain. They also reasoned, by extension, that schizophrenia might result in part from excessive action of dopamine. In other words, reducing dopamine in the brain might account for both the drugs' therapeutic effects and their adverse side effects.

How would this work? How could a drug produce both undesirable and beneficial effects? It depends on where in the brain the drug acts.

When neurons release dopamine into a synapse, the dopamine ordinarily binds to receptors on target neurons. If those receptors are blocked by antipsychotics, the action of dopamine is attenuated. As it turns out, many typical antipsychotics act by blocking dopamine receptors. This finding bolstered the idea that either excessive dopamine production or an excessive number of dopamine receptors is an important factor in causing schizophrenia. It also supported the idea emerging from studies of Parkinson's disease that dopamine deficiency causes abnormal movement. Thus, understanding the role that dopamine plays in schizophrenia also taught us a bit more about the normal functioning of this neurotransmitter.

Most dopamine-producing neurons are located in two clusters in the midbrain: the ventral tegmental area and the substantia nigra. The axons that extend outward from these two clusters of neurons form the neural circuits known as the *dopaminergic pathways*. Two of these dopaminergic pathways—the *mesolimbic pathway* and the *nigrostriatal pathway*—are the neural pathways primarily affected in schizophrenia and are therefore the most important ones to examine in looking for treatments (fig. 4.3).

The mesolimbic pathway extends from the ventral tegmental area to parts of the prefrontal cortex, hippocampus, amygdala, and nucleus accumbens. These regions are important for thought, memory, emotion, and behavior—the mental functions that are adversely affected by schizophrenia. The nigrostriatal pathway begins in the substantia nigra and extends to the dorsal striatum, a region of the brain that is involved with spatial and motor functions. This is the pathway that degenerates in Parkinson's disease. Antipsychotic drugs act on both pathways, which explains how they can produce both therapeutic effects and adverse side effects.

To test the validity of the idea that typical antipsychotics block dopamine receptors, scientists had to identify the specific dopamine recep-

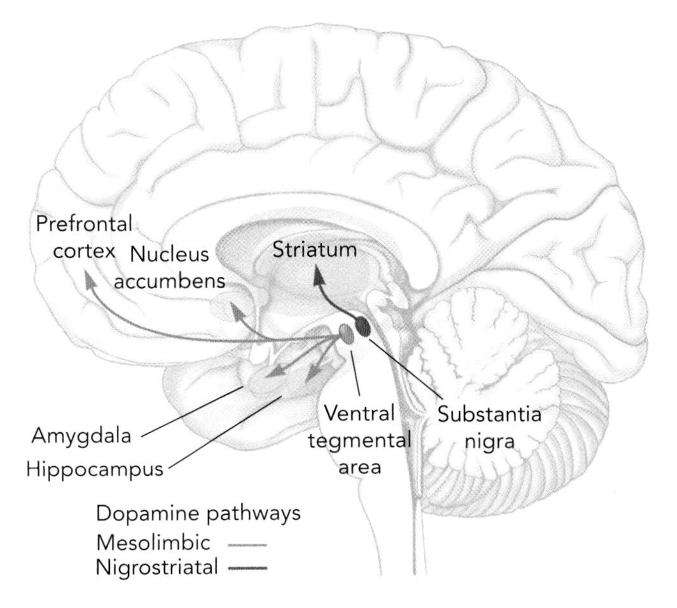

Figure 4.3. The two dopaminergic pathways affected by antipsychotic drugs: the mesolimbic pathway and the nigrostriatal pathway. Dopamine-producing neurons are concentrated in the ventral tegmental area, which transmits dopamine along the mesolimbic pathway, and in the substantia nigra, which sends dopamine along the nigrostriatal pathway.

tors on which the drugs exert their effect. There are five major known types of dopamine receptors, D1 to D5. Typical antipsychotic drugs were found to have a high affinity for the D2 receptor; atypical antipsychotics have a lower affinity for this receptor.

D2 receptors are normally present in particularly large numbers in the striatum and to a lesser extent in the amygdala, the hippocampus, and parts of the cerebral cortex. Research suggests that wholesale blocking of D2 receptors in the nigrostriatal pathway results in too little dopamine in regions of the striatum that require adequate dopamine for normal movement. This explains the Parkinson's-like effects of typical antipsychotics. Atypical antipsychotics also block the D2 receptors in the striatum, but because these drugs have a lower affinity for D2 receptors, they block fewer of them, thus leaving movement intact.

Another way in which atypical antipsychotics differ from typical ones is that their affinities are more diverse. Atypical antipsychotics bind to D4 dopamine receptors and to receptors for other modulatory neurotransmitters as well, notably serotonin and histamine. This diversity of action raises the possibility that schizophrenia involves abnormalities in serotonergic and histaminergic pathways as well as in dopaminergic pathways.

EARLY INTERVENTION

One key to improved treatment of any medical disorder is early intervention. Scientists have successfully identified high-risk lifestyles for a heart attack and have developed interventions to prevent them. Why not do the same for schizophrenia?

We know that genetic and environmental factors act on the developing brain before birth and in early childhood to increase the risk of schizophrenia, and we may eventually be able to pinpoint them and intervene before the disease manifests itself years later. One genetic variation that acts on the developing brain has already been identified, as we shall see later. In addition, computerized brain imaging can sometimes indicate areas of increased dopamine activity, which might serve as a biomarker of the disease before psychosis develops.

As we have seen, the first psychotic episode of schizophrenia is usually triggered in late adolescence or early adulthood, when the stresses of daily life can prove too heavy a burden to bear. If treatment is begun immediately, young people can usually be stabilized. All too often, however, they don't seek treatment until after they have been sick for several years. In addition, if a person with schizophrenia stops taking medication, the regulation of dopaminergic pathways and other neural circuits will be disrupted, and he or she will begin to experience symptoms again.

The most promising preemptive treatment thus far is to provide cognitive psychotherapy to adolescents and young adults who exhibit early signs of schizophrenia, in what is known as the *prodromal phase*. These signs, which precede the first psychotic episode, are unfortunately a bit vague. A young person may be slightly depressed, not handling stress as well as usual, or feeling less inhibited than usual—often saying out loud what he or she is thinking. As we know, major psychiatric disorders are often characterized by exaggerations of everyday behavior, so initial, subtle changes can be difficult to recognize.

Preemptive treatments are designed to help young people build up the cognitive capacity and executive functions of the prefrontal cortex that regulate their ability to control their behavior. This will improve their ability to manage day-to-day stress and organize their lives more effectively, thereby reducing the likelihood that they will have a psychotic episode.

PREDISPOSING ANATOMICAL ABNORMALITIES

During pregnancy, environmental factors, such as nutritional deficits, infections, or exposure to stress or toxins, may interact with genes to increase the risk that the fetus will develop abnormally functioning dopaminergic pathways. Malfunctioning pathways set the stage for developing schizophrenia years later, when the brain of the adolescent responds to the stresses of everyday life by generating excessive dopamine.

The same adverse environmental events or situations during pregnancy may also affect the way certain circuits in the prefrontal cortex develop, circuits that mediate the thinking and executive functions of the brain. Abnormalities in these neural circuits result in the cognitive symptoms that people with schizophrenia experience, notably a disturbance of working memory.

Think of working memory as the ability to remember, for a short period of time, the information you need to guide your thoughts or behavior. Right now, you are using your working memory to keep in mind the points you just read so that the next thing you read will follow logically. Impaired working memory would make this difficult, just as it would make it hard for you to plan your day or hold a job.

Working memory develops from childhood through the late teens, getting progressively better over time. At age seven, children who will be diagnosed with schizophrenia ten or fifteen years later have normal working memory. But by age thirteen, their working memory has fallen well below where it should be at that stage of development. A key component of working memory is the pyramidal neurons of the prefrontal cortex, so called because the cell body of these neurons is shaped roughly like a triangle. In every other respect these cells are like other neurons, both structurally and functionally.

As we have seen, neurons send information outward along the axon, which forms synaptic connections with a target cell's dendrites. Most of a pyramidal neuron's synapses are located on small protrusions from the dendrites called *dendritic spines*. The number of dendritic spines on a neuron is a rough measure of the amount and richness of the information it receives.

Dendritic spines begin to form on pyramidal neurons during the third trimester of pregnancy. From then through the first few years of life, the number of dendritic spines, and the number of synapses on them, expands rapidly. In fact, a three-year-old's brain contains twice as

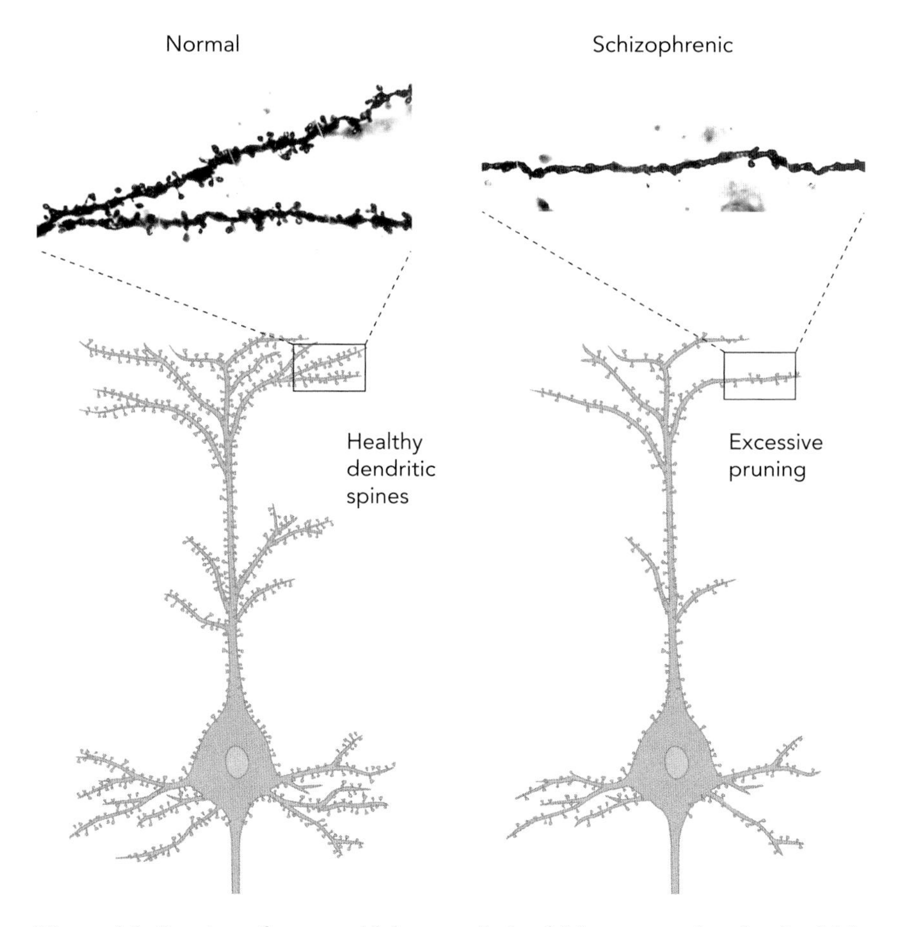

Figure 4.4. Pruning of a pyramidal neuron's dendritic outgrowth—the dendritic spines in the normal brain and the brain of someone with schizophrenia

many synapses as an adult's brain. Beginning at about puberty, synaptic pruning removes the dendritic spines that the brain isn't using, including spines that aren't actually helping working memory. Synaptic pruning becomes particularly active during adolescence and early adulthood.

In schizophrenia, synaptic pruning appears to go haywire during adolescence, snipping off far too many dendritic spines (fig. 4.4). Consequently, the pyramidal neurons are left with too few synaptic connections in the prefrontal cortex to form the robust neural circuits we need for an adequate working memory and other complex cognitive functions. This excessive-pruning hypothesis for schizophrenia, first proposed by Irwin Feinberg, now at the University of California, Davis,[2] has been documented by David Lewis and Jill Glausier at the University of Pittsburgh.[3] A similar defect is thought to affect pyramidal neurons located in the hippocampus of people with schizophrenia, which would adversely affect memory.

Since synaptic pruning is designed to rid the brain of unused dendrites, Lewis reasoned that excessive pruning might be the result of not having enough dendrites in play—that is, something might be preventing the pyramidal neurons from receiving enough sensory signals to keep the dendritic spines busy and functional. The likely culprit in this case would be the thalamus, the part of the brain that is supposed to relay sensory signals to the prefrontal cortex. If the thalamus has fallen down on the job, it might be because the thalamus itself has lost cells. Indeed, some studies have found that the thalamus is smaller than normal in people with schizophrenia.

Thus, schizophrenia presents quite a different problem than depression or bipolar disorder. As we saw in chapter 3, those disorders result from a *functional* defect, in which properly built neural circuits fail to work correctly. Such defects can often be reversed. Schizophrenia, like autism spectrum disorders, involves an *anatomical* defect, in which certain neural circuits fail to develop correctly. To remedy these anatomical defects in schizophrenia, scientists will have to think of some way to either intervene in synaptic pruning during development or create compounds that stimulate the growth of new spines later on.

Schizophrenia is characterized by other anatomical abnormalities as well. These include a thinning of layers of gray matter in the temporal and parietal regions of the cortex and in the hippocampus, as well as dilation of the lateral ventricles, the hollow spaces that carry the cerebrospinal fluid. Enlargement of the lateral ventricles probably results secondarily from the loss of gray matter in the cortex. Like excessive synaptic pruning, these brain abnormalities appear early in life, which suggests that they contribute to the development of schizophrenia. The existence of anatomical abnormalities and their parallel to the emergence of cognitive symptoms have strengthened the longstanding belief that the cognitive symptoms of schizophrenia emerge from abnormal functioning of the gray matter of the cerebral cortex.

THE GENETICS OF SCHIZOPHRENIA

If you had an identical twin with schizophrenia, you would have about a 50-50 chance of developing the disease, regardless of whether the two of you were raised together or apart. That risk of developing schizophrenia is much higher than the 1 in 100 risk for the general population. The twin data tell us two things: first, schizophrenia has a strong genetic component, regardless of environment; and second, those genes can't be acting alone, because the risk isn't 100 percent. Genes and the environment must interact to cause the disease (fig. 4.5).

In recent years, a collaboration involving many scientists and tens of thousands of schizophrenic patients and their families set out to understand that genetic risk. They wanted to know what genes contribute to the brain abnormalities of people with schizophrenia and what sorts of functions those genes mediate.[4] They found that even though symptoms of the disease don't appear until the late teens, many of the genes involved in schizophrenia act on the developing brain before birth. This finding is consistent with the fact that people are vulnerable to environmental risk factors early in life, even though they do not manifest signs of disease until much later.

Scientists have recently come to appreciate that genetic variations that contribute to complex disorders such as autism, schizophrenia, or bipolar disorder may be either common or rare. A common variation is one that was introduced into the human genome many generations ago and is now present in more than 1 percent of the world's population; such variations are called *polymorphisms*. Rare variations, or mutations,

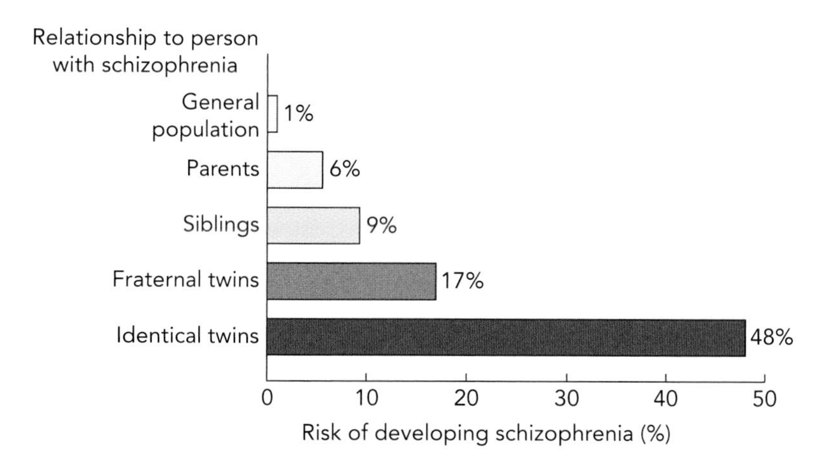

Figure 4.5. The genetic risk of developing schizophrenia. As this graph shows, the general population has a 1 in 100, or 1 percent, risk of developing schizophrenia, whereas relatives of someone with the disorder have a higher risk, reaching almost 50 percent in identical twins.

occur in less than 1 percent of the world's population. Either type of variation can contribute to the likelihood of having a disease or developmental disorder. Each type of variation can predispose a person to schizophrenia.

The rare variant mechanism of disease illustrates that rare mutations in a person's genome greatly increase that person's risk of developing a relatively common disorder. As we saw in chapter 2, a rare change in the structure of a chromosome, known as a copy number variation, can markedly increase the risk of autism spectrum disorders. The same is true of schizophrenia—in fact, the same copy number variation on chromosome 7 that increases the risk of autism spectrum disorders also increases the risk of schizophrenia. Moreover, as is the case with autism spectrum disorders, rare de novo mutations in DNA—mutations that occur spontaneously in the sperm of the father—increase the risk of schizophrenia and bipolar disorder. Because the sperm of older men continue to divide and the older sperm undergo more frequent mutations, older fathers are more likely than younger fathers to have children who develop schizophrenia.

The common variant mechanism of disease illustrates that both schizophrenia and autism spectrum disorders result when many common polymorphisms of a number of different genes act together to increase risk. Unlike the rare mutation, which exerts an outsize effect on risk, each of these common variants exerts only a very small effect. The strongest evidence for the common variant mechanism comes from the collaborative study of schizophrenia. These scientists have studied associations between schizophrenia and millions of common variants in the genomes of tens of thousands of individuals. Approximately one hundred gene variants related to schizophrenia have already been found. In this respect, the genetics of schizophrenia closely mirrors that of other common medical conditions, such as diabetes, heart disease, stroke, and autoimmune disorders.

For a while, the rare variant and common variant mechanisms of disease were thought to be mutually exclusive, but recent studies of autism, schizophrenia, and bipolar disorder suggest that each disorder has an underlying genetic risk, quite apart from any rare genetic variation caused by copy number variations or de novo mutations (chapter 1, table 1). The underlying risk for schizophrenia, for example, is 1 percent, or 1 in 100 people in the general population. The relative contribution of rare and common genetic variations to underlying risk is somewhat different for each disorder, but certain characteristics seem to be universal. Common variations, each of which carries a small risk, contribute to the disorder in relatively large numbers of people, whereas rare mutations, each of which carries a larger risk, typically contribute to the disorder in fewer than 1 in 100 affected individuals.

Perhaps the most surprising recent finding uncovered by the large collaborative effort on the genetics of schizophrenia is that some of the same genes that create a risk for schizophrenia also create a risk for bipolar disorder. What's more, a different group of genes that creates a risk for schizophrenia also creates a risk for autism spectrum disorders.

So here we have three different diagnoses—autism, schizophrenia, and bipolar disorder—sharing genetic variants. This overlap suggests that the three disorders have other features in common early in life.

DELETED GENES

One out of every four thousand babies is born with a piece of chromosome 22 missing from its genome. The amount of DNA that is missing can vary, but it usually involves about 3 million DNA building blocks, known as *base pairs*, resulting in the loss of between thirty and forty genes. Because the missing DNA is from a region near the middle of the chromosome, at a location designated q11, people with the deletion are said to suffer from 22q11 deletion syndrome.

The syndrome can cause highly variable symptoms. Almost everyone with the deletion has abnormalities of the head and face, such as cleft lip or cleft palate, and over half have cardiovascular disorders. They also display cognitive deficits that range from impaired working memory and executive function, as well as mild learning disabilities, to mental retardation. About 30 percent of adults with the syndrome are diagnosed with psychiatric disorders, including bipolar disorder and anxiety disorders. But schizophrenia is by far the most prevalent disorder. In fact, the risk of schizophrenia in a person with 22q11 deletion syndrome is twenty to twenty-five times greater than the risk of schizophrenia in the general population.

To find out which genes might be responsible for the various medical problems associated with the syndrome, scientists looked for an animal in which to model the deletion. It turns out that a segment of DNA in chromosome 16 in the mouse has almost all of the genes present in the q11 region of chromosome 22 in humans. By deleting a different section of the region from different mice, the scientists were able to generate several mouse models of the human syndrome.

The models revealed that the loss of a transcription factor—a protein involved in gene expression—is responsible for many of the non-psychiatric medical conditions suffered by humans, including cleft palate and some heart defects. Many scientists are now using mouse models to determine which specific genes within the 22q11 region, when missing, contribute to schizophrenia. Considering the prevalence of schizophrenia in people with this deletion, the scientists have a good chance of identifying those genes.

In 1990 David St. Clair, then at the University of Edinburgh, and his colleagues described a Scottish family with a high prevalence of mental illness.[5] Thirty-four members of the family carry what is known as a balanced autosomal translocation. This means that pieces of two different non-sex chromosomes have broken off and switched places. Of the thirty-four family members who carry this particular translocation, five were diagnosed with schizophrenia or schizoaffective disorder (schizophrenia plus mania and/or depression) and seven with depression.

The researchers identified two genes that are disrupted by the translocation: *DISC1* (*disruption in schizophrenia 1*) and *DISC2* (*disruption in schizophrenia 2*). Although this particular translocation has been found in only one family, that family's unusually high incidence of psychiatric disorders suggests that *these two genes*, and other genes close to where the chromosomes broke, may be responsible for psychotic symptoms in schizophrenia and in mood disorders. Two separate groups of researchers found another genetic clue: some polymorphisms in the *DISC1* gene occur together frequently and seem to contribute to the risk of schizophrenia.[6] So far, studies have been focused on the *DISC1* gene because the *DISC2* gene does not produce a protein; however, the *DISC2* gene is thought to play a role in regulating the *DISC1* gene.

Numerous studies in fruit flies and mice have found that *DISC1* affects a variety of cell functions throughout the brain, including intracellular signaling and gene expression. *DISC1* is particularly important in the developing brain: it helps neurons migrate to their proper location in the fetal brain, to position themselves, and to differentiate into various cell types. Disruption of the *DISC1* gene compromises its ability to perform these critical developmental functions.

Taken together, the mouse models show quite clearly that the disrupted functions of the *DISC1* gene lead to deficits typical of schizophrenia. In addition, all of the models show changes in brain structure that are similar to the ones observed in people with schizophrenia. Brain-imaging studies of one model, for example, show the enlarged lateral ventricles and smaller cortex seen in people with schizophrenia. Another model shows that disrupting the gene's function soon after birth produces abnormal behavior in the adult animal. The apparent role of the *DISC1* gene in schizophrenia and the findings in mice are consistent with the idea of schizophrenia as a disorder of brain development.

GENES AND EXCESSIVE SYNAPTIC PRUNING

Normal synaptic pruning, in which the brain trims unneeded connections between neurons, is extremely active during adolescence and early adulthood and takes place primarily in the prefrontal cortex. As we have seen, people with schizophrenia have fewer synapses in this area of the brain than unaffected people do, so researchers have long suspected that synaptic pruning is excessive in schizophrenia.

Recently, Steven McCarroll, Beth Stevens, Aswin Sekar, and their colleagues at Harvard Medical School provided further evidence in support of this idea. They also described how and why pruning may go wrong, and they have identified the gene responsible.[7]

The researchers focused on a particular region of the human genome, a locus called the major histocompatibility complex (MHC). This complex of genes on chromosome 6 encodes proteins that are essential for recognizing foreign molecules, a critical step in the body's immune response. The MHC locus, which had been strongly associated with schizophrenia in previous genetic studies, contains a gene called *C4*. The activity of the *C4* gene—that is, its level of expression—varies significantly among individuals. The researchers wanted to find out how variations in the *C4* gene are related to its level of expression and whether its level of expression is related to schizophrenia.

McCarroll, Stevens, Sekar, and their colleagues analyzed the genomes of more than sixty-four thousand people with and without schizophrenia and found that the people with schizophrenia were more likely to carry a particular variant of the *C4* gene known as *C4-A*. This finding suggested that *C4-A* may increase the risk of schizophrenia.

Earlier studies had found that proteins produced by genes in the MHC locus play a role in immunity and are involved in synaptic pruning during normal development. This raised a critical question: What exactly is the role of the protein product produced by the *C4-A* gene? To answer the question, scientists bred mice without the gene. They observed less-than-normal synaptic pruning in these mice, indicating that the protein's role is to promote pruning and suggesting that too much of the protein leads to excessive pruning. In studies of these mice McCarroll, Stevens, Sekar, and their colleagues also found that during normal development the C4-A protein "tags" the synapses to be pruned. The more active the *C4* gene is, the more synapses are deleted.

Together, these studies suggest that overexpression of the *C4-A* variant leads to excessive synaptic pruning. Excessive pruning during late adolescence and early adulthood—when normal synaptic pruning kicks into overdrive—changes the anatomy of the brain and accounts for both the late onset of schizophrenia and the thinner prefrontal cortex of people with the disorder.

Carrying a gene variant that facilitates aggressive pruning is not enough in itself to cause schizophrenia; many other factors are also at work. But in a small subgroup of people, one specific gene—the *C4-A* gene—gives rise to anatomical changes that lead to schizophrenia. Thus, McCarroll, Stevens, Sekar, and their colleagues have given us the first real inroad into the etiology of schizophrenia, an inroad that may eventually lead to new treatments. Moreover, important studies such as these inspire other researchers who are trying to use genetics to advance our understanding of psychiatric disorders.[8]

MODELING THE COGNITIVE SYMPTOMS OF SCHIZOPHRENIA

Earlier, we learned that excessive dopamine production may contribute to the development of schizophrenia and that antipsychotic drugs produce their effects by blocking dopamine receptors in the mesolimbic pathway. We also learned that brain-imaging studies have found both more dopamine and more D2 receptors in the striatum of people with schizophrenia. Moreover, in at least some people, the greater-than-normal number of D2 receptors may be determined genetically. In light of these findings, Eleanor Simpson, Christoph Kellendonk, and I set out to determine whether an excessive number of D2 receptors in the striatum causes the cognitive symptoms of schizophrenia.[9]

To do so, we created a mouse model containing a human gene that overexpresses D2 receptors in the striatum. We found that this transferred gene, or transgene, impairs in the mouse the same cognitive processes that are affected in people with schizophrenia. In addition, the mouse lacked motivation, a deficit that is characteristic of the negative symptoms of schizophrenia. But the most interesting result was that whereas the motivational deficits disappeared once the transgene was switched off, the cognitive deficits did not—they persisted long afterward. In fact, the action of the transgene during the period of prenatal development alone was sufficient to cause cognitive deficits in adulthood.

These findings suggest three important new ideas.

First, excessive action of dopamine in the mesolimbic pathway, resulting from an overabundance of D2 receptors, could be the main cause of schizophrenia's cognitive symptoms—because this pathway connects to the prefrontal cortex, the site of the cognitive symptoms. Second, antipsychotics that block D2 receptors ease the positive symptoms of schizophrenia but have little, if any, beneficial effect on the cognitive symptoms. Why? Because this medication is given too late in development—long after irreversible changes have taken place. Third, because cognitive and negative symptoms are strongly correlated in people with schizophrenia, they may be caused by some of the same factors.

All of these remarkable manipulations—creating deletions, inserting transgenes, and increasing the number of D2 receptors in mice—are just some of the many tools scientists are now using to discover the causes of schizophrenia, depression, and bipolar disorder. In a larger sense, these manipulations are beginning to give us some insights into the relationship of brain science to cognitive psychology, of the relationship of brain to mind.

LOOKING AHEAD

Before moving on to considerations of other brain disorders, it is worth reexamining some of the important contributions research has made to our understanding of the healthy brain from studies of autism spectrum disorders, mood disorders, and schizophrenia.

The importance of brain imaging can scarcely be overestimated. Our understanding of where and how psychiatric and autism spectrum disorders affect the brain has advanced hand in hand with advances in imaging technology. And because imaging studies generally compare the brains of people with and without a particular mental disorder, they have given us additional insights into the healthy human brain as well. Imaging has advanced to the point where it can show us what regions, and sometimes even what neural circuits within those regions, are essential for normal functioning.

Imaging has also confirmed that psychotherapy is a biological treatment—that it physically changes the brain, as drugs do. Imaging has even predicted, in some cases of depression, which patients are best treated with drugs, with psychotherapy, or with both.

We have also seen how critical insights into the nature of depression and schizophrenia came about by accident, when drugs designed to treat another disorder were observed to have an effect on patients with these brain disorders. Subsequent research on how the drugs act in the brain revealed important biochemical underpinnings of depression and schizophrenia and led to better treatments for people with these disorders.

Advances in genetics are uncovering how genetic variations—whether common or rare—create a risk of developing complex brain disorders. Particularly fascinating is the discovery of shared genes that operate in schizophrenia and bipolar disorder, and in schizophrenia and autism spectrum disorders. Such insights into the molecular nature of depression and schizophrenia have also improved our understanding of normal mood and of organized thought.

Finally, we are again reminded of how much we owe to animal models of disease. Genetic studies of social behavior in animals have shown that some of the same genes that contribute to social behavior in animal models also contribute to our own social behavior; mutations in those genes may therefore be involved in autism spectrum disorders. Recent studies of schizophrenia, in particular, have relied heavily on mouse models for vital clues to the causes of this disorder of thought and volition.

In a larger sense, the studies of autism, depression, bipolar disorder, and schizophrenia—and the brain functions they affect—have yielded profound insights into the nature of our mind and our sense of self. These insights are informing a new understanding of human nature and thereby contributing to the emergence of a new humanism.

5

MEMORY, THE STOREHOUSE OF THE SELF: DEMENTIA

Learning and memory are two of the most wondrous capabilities of our mind. Learning is the process whereby we acquire new knowledge about the world, and memory is the process whereby we retain that knowledge over time. Most of our knowledge about the world and most of our skills are not inherent but learned, built up over a lifetime. As a result, we are who we are in good measure because of what we have learned and what we remember.

Memory is part and parcel of every brain function, from perception to action. Our brain creates, stores, and revises memories, constantly using them to make sense of the world. We depend on memory for thinking, learning, decision making, and interacting with other people. When memory is disrupted, these essential mental faculties suffer. Thus, memory is the glue that holds our mental life together. Without its unifying force, our consciousness would be broken into as many fragments as there are seconds in the day.

No wonder we worry about the continued reliability of our memory.

We have seen that disturbances of memory accompany depression and schizophrenia, but what about loss of memory per se? Is memory loss inevitable as we age? Is normal age-related memory loss different from Alzheimer's disease and other disorders that affect memory?

This chapter first describes what we know about memory, including how we learn and how our brain stores what we have learned as memory.

It then considers the aging brain and three neurological disorders that affect memory: age-related memory loss, Alzheimer's disease, and frontotemporal dementia. Both Alzheimer's and frontotemporal dementia, as well as Parkinson's disease and Huntington's disease, which we will discuss in chapter 7, are thought to be caused in part by faulty protein folding. But before exploring the aging brain and protein folding, let's touch on different types of memories, how they are created, and where in the brain they are stored.

THE SEARCH FOR MEMORY

Memory is a complex mental function—so complex, in fact, that scientists initially questioned whether it was even possible for memory to be stored in a particular region of the brain. Many thought it was not. However, as we saw in chapter 1, the noted Canadian neurosurgeon Wilder Penfield made an astonishing discovery in the 1930s. When he stimulated the temporal lobe of his epileptic patients prior to surgery (fig. 5.1), some of them seemed to be recalling memories, such as a lullaby their mother used to sing to them or the recollection of a dog chasing a cat.

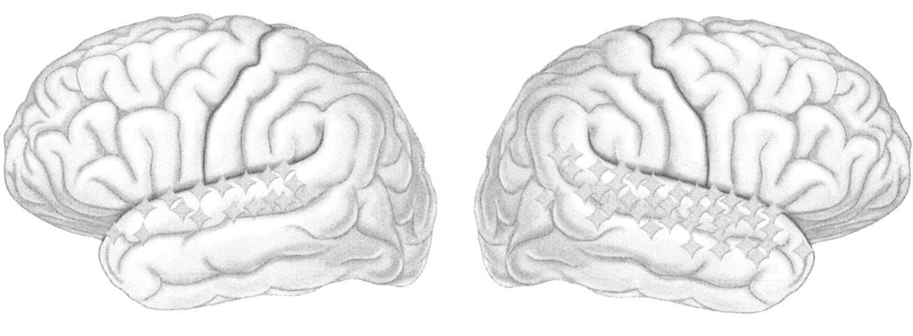

Figure 5.1. Stimulation points (diamond shapes) on the temporal lobe that elicit auditory memory in the left and right hemispheres of the brain

Penfield had earlier outlined sensory and motor maps of brain function, but memory was a different, more complicated matter. He called in Brenda Milner, an extraordinarily gifted young cognitive psychologist on the staff of the Montreal Neurological Institute, and together they investigated the temporal lobe, particularly its medial (inner) surface, and its role in memory.

One day, Penfield received a telephone call from William Scoville, a neurosurgeon working in New Haven, Connecticut, who had recently operated on a man suffering from severe seizures. That man was H.M. (fig. 5.2), who became one of the most important patients in the history of neuroscience.

Figure 5.2. H.M.

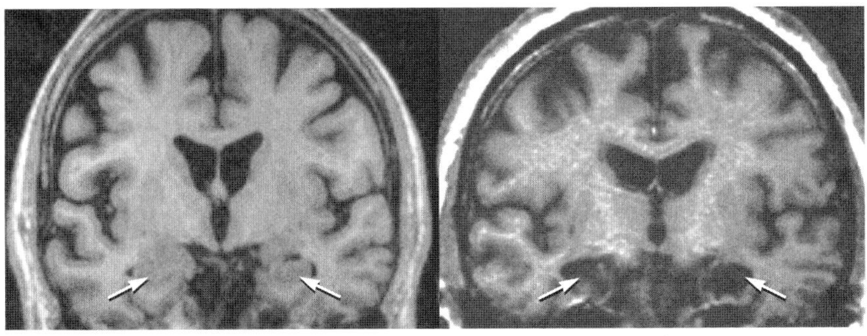

Figure 5.3. Comparison of an intact brain and H.M.'s brain, with part of the medial region of both temporal lobes removed (arrows)

H.M. had been run over by a bicycle rider when he was nine years old. The resulting head injury led to epilepsy. By age sixteen, he had begun having major convulsions. He was treated with the maximum doses of the anticonvulsant medication available at that time, but the medication didn't help him. Although he was bright, he had great difficulty finishing high school and keeping a job because of his frequent seizures. Eventually, H.M. went to Scoville for help. Scoville inferred that H.M. suffered from scarring of the hippocampal structures lying deep within the temporal lobes. He therefore removed a part of the medial region of the temporal lobe—including the hippocampus—on both sides of H.M.'s brain (fig. 5.3).

The operation essentially cured H.M.'s epilepsy, but it left him with severe memory disturbance. Although he remained the polite, gentle, calm, and pleasant young man he had always been, he had lost the ability to form any new long-term memories. He remembered people he had known for many years before the operation, but he did not remember anyone he had met since the operation. He couldn't even learn how to get to the bathroom in the hospital. Scoville invited Milner to study H.M., and she ended up working with him for twenty years. Yet each time she walked into the room, it was as if H.M. were meeting her for the first time.

For a long time Milner thought that H.M.'s memory deficit applied to all areas of knowledge. Then she made a remarkable discovery. She asked H.M. to trace the outline of a star while looking at his hand, his pencil, and the paper in a mirror. Everyone who tries this tracing task makes errors on the first day, drawing outside the line of the star and having to adjust back in, but people with normal memory improve to almost perfect performance by the third day. If H.M.'s memory loss applied to all areas of knowledge, he should show no such improvement. Yet after three days, and despite having no memory of practicing the task or of having seen Milner before, H.M. had learned this motor task as well as anyone else (fig. 5.4).

Because H.M. could not remember having practiced, scientists speculated that motor learning, unlike every other form of learning, must involve a special form of memory. It must be mediated by other systems in the brain.

Neuroscientists thought this for a very long time—until Larry Squire

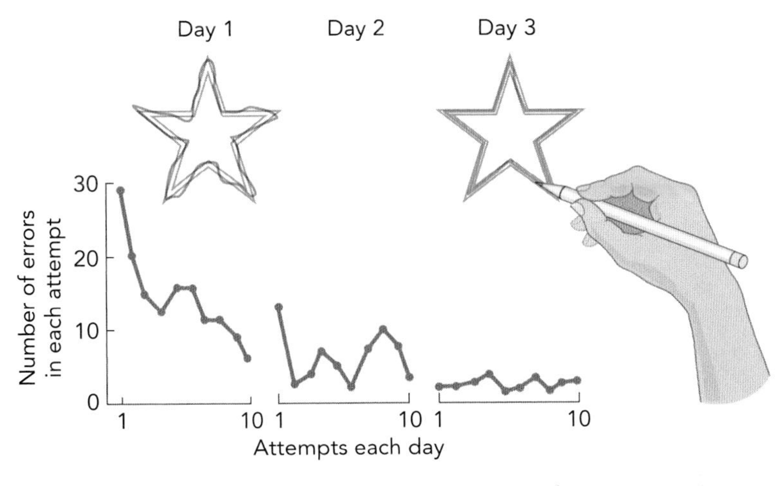

Figure 5.4. Learning a motor task

at the University of California, San Diego, found that people with damage to the medial region of both temporal lobes (the same areas removed in H.M.) can learn more than motor skills. Their capability for language is normal, and they can perform a whole family of learned perceptual skills as well, such as reading mirror-reversed print. They can also acquire habits and other simple forms of learning. If this range of learning capabilities remained, Squire reasoned, then maybe these people were relying on a different kind of memory system.[1]

Squire came to realize that there are two major memory systems in the brain. One is *explicit*, or *declarative, memory*, which allows us to consciously remember people, places, and objects. This is what we mean when we refer to "memory" in everyday language. It reflects our conscious ability to remember facts and events. Explicit memory relies on the medial region of the temporal lobe, which explains why H.M. could no longer remember new facts or people or the events of his passing days.

The second type of memory, the memory that Squire identified, is *implicit*, or *non-declarative, memory*, which our brain uses for motor and perceptual skills that we do automatically, like driving a car or using correct grammar. When you speak, you are usually not conscious of using correct grammar—you just speak. What makes implicit memory so mysterious—and the reason we rarely pay attention to it—is that it is largely unconscious. Our performance of a task improves as a result of experience, but we are not aware of it, nor do we have the sense of using memory when we perform the task. In fact, studies show that performance on implicit tasks can actually be impaired when we consciously contemplate the action.

Not surprisingly, implicit memory depends on different brain systems than explicit memory does. Rather than relying on higher, cognitive regions such as the medial region of the temporal lobe, implicit memory depends more on regions of the brain that respond to stimuli, for example the amygdala, the cerebellum, and the basal ganglia, or, in the simplest instances, the reflex pathways themselves.

A particularly important subclass of implicit memory is evident in memory that is associated with conditioning. Aristotle was the first person to suggest that certain types of learning require the association of ideas. For example, whenever you see a tree covered with lights you think of Christmas. This notion was elaborated and formalized by the British Empiricists John Locke, David Hume, and John Stuart Mill, the forefathers of modern psychology.

In 1910 the Russian physiologist Ivan Pavlov extended this idea one critical step further. In earlier studies of dogs, he had noticed that the animals began to salivate when he entered the room, even when he was not carrying their food. In other words, the dogs had learned to associate a neutral stimulus (his entering the room) with a positive stimulus (food). Pavlov called the neutral stimulus a *conditioned* stimulus and the positive stimulus an *unconditioned* stimulus—and he called this form of associative learning *conditioning.*

Based on his observation, Pavlov designed an experiment to see whether a dog would learn to salivate in response to any signal that predicted the arrival of food. He rang a bell and then gave the dog food. At first, ringing the bell produced no response. After several instances of pairing the sound of the bell with food, however, the dog salivated in response to the ringing of the bell, even when no food was forthcoming.

Pavlov's work had an extraordinary impact on psychology: it marked a decisive shift toward a behavioral concept of learning. To Pavlov, learning involved not only an association between ideas but also an association between a stimulus and behavior. This made learning amenable to experimental analysis: responses to stimuli could be measured objectively, and the parameters of a response could be specified or even modified.

Squire's discovery that memory is not a unitary function—that different kinds of memories are processed in different ways and stored in different regions of the brain—was a major advance in our understanding of memory and in our understanding of the brain, but it inevitably raised a new set of questions. How do neurons store these different types of memories? Are different cells responsible for implicit and explicit memories? If so, do they operate differently?

MEMORY AND THE STRENGTH OF SYNAPTIC CONNECTIONS

Early studies assumed that it takes a fairly complex neural circuit to form and store a memory of what we have learned. However, my colleagues and I at Columbia University and Jack Byrne, one of my former students, now at the University of Texas Health Science Center at Houston, encountered a mechanism of associative learning in the invertebrate marine snail *Aplysia* that does not require a complex neural circuit.[2] *Aplysia* has an important defensive reflex that is mediated by the connections between a small number of sensory neurons and motor neurons. Learning leads to the activation of modulatory neurons, which strengthen the connections between the sensory and motor neurons. My colleagues and I found that this mechanism contributes to implicit learning of classical conditioning in invertebrate animals. It also operates in the amygdala, the brain structure in mammals that is critical for implicit learning of emotion, particularly fear.

Another person who challenged the idea that a complex neural circuit is required for learning was the Canadian psychologist Donald Hebb. Hebb suggested that associative learning could be produced by the simple interaction of two neurons: if neuron A repeatedly stimulates neuron B to fire an action potential—the electrical impulse that travels down the axon to the synapse—a change will take place in one or both of those cells. That change strengthens the synaptic connection between the two neurons. The strengthened connection creates and stores, for a short time, a memory of the interaction.[3] Two researchers working at the University of Göteborg in Sweden, Holger Wigström and Bengt Gustafsson, later provided the first evidence suggesting that Hebb's mechanism might be at work in the formation of explicit memory in the hippocampus.[4]

Both implicit and explicit memory can be stored in the short term, for minutes, and in the long term, for days, weeks, or even longer. Each form of memory storage requires specific changes in the brain. Short-term memory results from strengthening existing synaptic connections, making them function better, whereas long-term memory results from the growth of new synapses. Put another way, long-term memory leads to anatomical changes in the brain, whereas short-term memory does not. When synaptic connections weaken or disappear over time, memory fades or is lost.

MEMORY AND THE AGING BRAIN

Thanks to a broad range of medical advances, the average American born today is expected to live for about eighty years, in contrast to only fifty years in 1900. For many elderly Americans, however, this welcome increase in life expectancy is marred by a deterioration in cognitive abilities, particularly memory (fig. 5.5).

Some weakening of memory, beginning around age forty, is normal. Until recently, however, it was not clear whether this age-related memory loss, also called *benign senescent forgetfulness*, is simply the early phase of Alzheimer's disease or a distinct entity in its own right. The answer to that question is not only a matter of considerable scientific interest, it is also a matter of enormous financial and emotional consequence to our society and its aging population.

Because implicit and explicit memory are controlled by different systems in the brain, aging affects them differently. Implicit memory is often well preserved in old age, even in the early stages of Alzheimer's disease. That's because the disease does not affect the amygdala, the cerebellum, or other areas important for implicit memory until quite late in its course. It also explains why people who are unable to recall the names of loved ones can still ride a bicycle, read a sentence, and play the piano. In contrast, explicit memory—the memory of facts and events—degrades early in people with Alzheimer's disease.

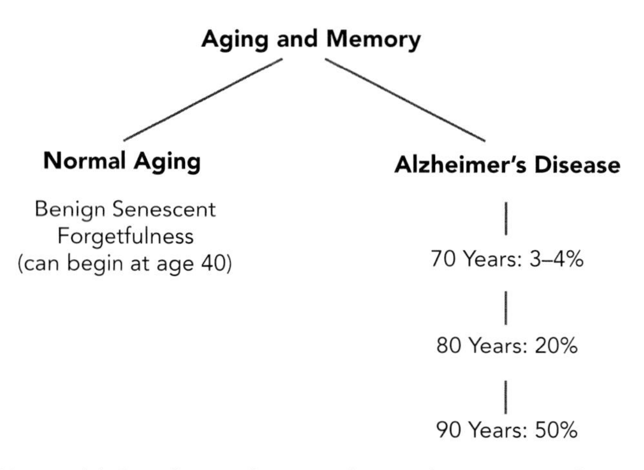

Figure 5.5. Prevalence of memory loss in the aging population

To find out whether Alzheimer's and age-related memory loss are biologically different, two groups of scientists at Columbia University, one led by Scott Small and one by me, compared three variables: namely, age of onset and progression of each disorder, regions of the brain involved, and molecular defects in each of the identified regions.

To compare age of onset and progression, my colleagues and I turned to mice.[5] Mice do not develop Alzheimer's disease, but we found that they do show an age-related memory loss that is centered in the hippocampus. This memory loss begins in midlife, as age-related memory loss appears to do in people. So in mice, at least, we could see that age-related memory loss exists as a separate entity, independent of Alzheimer's disease.

To find out what areas of the brain are involved in age-related memory loss and what areas are involved in Alzheimer's disease, Small and his group used brain imaging to study human volunteers ranging in age from thirty-eight to ninety. They found, as others had earlier, that Alzheimer's disease begins in the entorhinal cortex, but they also found that age-related memory loss involves the dentate gyrus, a structure within the hippocampus.[6]

Small's group and mine then collaborated to determine whether the dentate gyrus contains any molecular defects that the entorhinal cortex does not contain.[7] To do this, we examined at autopsy the brains of people between the ages of forty and ninety who did *not* have Alzheimer's disease. Using Affymetrix GeneChips, a technology that enabled us to analyze changes in the expression of as many as twenty-three thousand genes, we found nineteen gene *transcripts* that varied with the age of the volunteer. (Transcripts are the single-strand RNA molecules produced in the initial stage of gene expression.) The first and most dramatic change was in a gene called *RbAp48*. This gene became increasingly less active in the dentate gyrus of older volunteers, resulting in less RNA transcription and less synthesis of the RbAp48 protein. Moreover, the change occurred only in the dentate gyrus, not in any other area of the hippocampus or in the entorhinal cortex.

RbAp48 turned out to be an interesting protein. It is part of the CREB complex, a group of proteins that are critical for turning on gene expression required for the conversion of short-term memory to long-term memory.

Finally, Small and I returned to mice to see whether expression of the RbAp48 protein also drops off in the dentate gyrus of mice as they age. We found that it does—and once again, the decrease occurs only in the dentate gyrus. In addition, we found that knocking out the *RbAp48* gene caused young mice to perform as poorly on spatial tasks as old mice. Conversely, ramping up the expression of the *RbAp48* gene in old mice eliminated age-related memory loss, causing them to perform like young mice.

At this point a surprise emerged. Gerard Karsenty, a geneticist at Columbia University, had picked up on the discovery that bone is an endocrine organ and that it releases a hormone called osteocalcin. Karsenty found that osteocalcin acts on many organs of the body and also gets into the brain, where it promotes spatial memory and learning by influencing the production of serotonin, dopamine, GABA, and other neurotransmitters.[8]

Karsenty and I joined forces to examine whether osteocalcin also affects age-related memory loss.[9] My colleague Stylianos Kosmidis injected osteocalcin into the dentate gyrus of mice and found that it leads to increased PKA, CREB, and RbAp48—the proteins needed for memory formation. Mice that were not given the injections had fewer CREB and RbAp48 proteins. Interestingly, when we gave old mice osteocalcin, their performance on memory tasks such as novel object recognition—which had declined with age—improved. In fact, their memory matched that of young mice. Moreover, osteocalcin even improved the learning capabilities of young mice.[10]

These findings—that osteocalcin declines with age and that it can reverse age-related memory loss in mice—may provide another explanation for the beneficial effects of exercise on the aging human brain. We know that aging is associated with a decrease in bone mass and that the resulting decrease of osteocalcin contributes to age-related memory loss in mice, and possibly in us as well. We also know that vigorous exercise builds bone mass. Thus it is likely that osteocalcin released by the bones ameliorates age-related memory loss in people as well as mice.

Clearly, as these studies illustrate, age-related memory loss is a disorder that is distinct from Alzheimer's disease—it acts on different processes in a different region of the brain. Moreover, the Roman ideal of a sound mind in a sound body now appears to have a scientific basis.

This is good news for people with a normally aging brain. They can maintain crucial mental functions into old age, provided they eat healthfully, exercise, and interact with others. Just as we have learned to extend the life of the body, we must also extend the life of the mind. Fortunately, as we have seen, several avenues of research encourage us to think that diseases affecting memory may one day be preventable.

It is also important to note that many of the aspects of cognitive function that don't require memory mature quite well. Wisdom and perspective certainly increase with age. Anxiety tends to decrease. The challenge for all of us is to maximize the benefits of aging while doing our best to minimize the downside.

ALZHEIMER'S DISEASE

Aging seems to target particular areas in the brain, and, as we have seen, the hippocampus is one of the most vulnerable. Sometimes it is damaged by lack of blood flow or cell death, but it is often damaged by Alzheimer's disease.

Alzheimer's disease is characterized by deficits in recent memory. It results from the loss of synapses, the point of contact where neurons communicate. The brain can regrow synapses in the early stages of the disease, but in the later stages, neurons actually die. Our brain cannot regrow neurons, so this cell death results in permanent damage. Treatment for Alzheimer's is likely to be most effective early on, before extensive cell death, so neurologists are trying to develop functional brain imaging and other methods of identifying the disease as early as possible.

Scientists have begun to unravel the cascade of events underlying the symptoms of Alzheimer's disease. They have also learned a great deal about the molecular biology of the disease. Every detail added to that store of knowledge gives us another potential target for a drug, another possible way of halting the progress of this devastating disorder.

The discovery of Alzheimer's disease dates back to 1906, when Alois Alzheimer, a German psychiatrist and colleague of Emil Kraepelin, described the case of a fifty-one-year-old woman, Auguste D., who had become suddenly and irrationally jealous of her husband. Soon thereafter, she developed memory deficits and a progressive loss of cognitive abilities. In time, her memory became so impaired that she could no longer orient herself, even in her own home. She hid objects. She started to believe that people intended to murder her. She was admitted to a psychiatric clinic and died less than five years after the onset of her symptoms.

Alzheimer performed an autopsy on Auguste D. and found three specific alterations in the cerebral cortex that have since proven to be characteristic of the disease. First, her brain was shrunken and atrophied. Second, the outside of the nerve cells contained deposits of a dense material that formed what we now call *amyloid plaques*. Third, inside the neurons was an accumulation of tangled protein fibers that we now call *neurofibrillary tangles*. Because of the importance of this discovery, Kraepelin named the disorder after Alois Alzheimer.

Some of what a pathologist sees under a microscope at autopsy we can now see with brain imaging. Figure 5.6 shows the amyloid plaques and neurofibrillary tangles that are hallmarks of Alzheimer's disease. At first, scientists thought these abnormal protein aggregates were just by-products of the disease, but we now know that they are instrumental in causing it. One of the fascinating things about them is that they form ten to fifteen years *before* a person's memory or thinking has begun to change. If these structures could be detected when they first appear, it might be possible to prevent damage to the brain and to stop Alzheimer's disease in its tracks.

Plaques initially form in specific, restricted areas of the brain. One such site is the prefrontal cortex. As we learned earlier, this part of the brain is involved in attention, self-control, and problem solving. Tangles start in the hippocampus. Amyloid plaques and tangles in these two

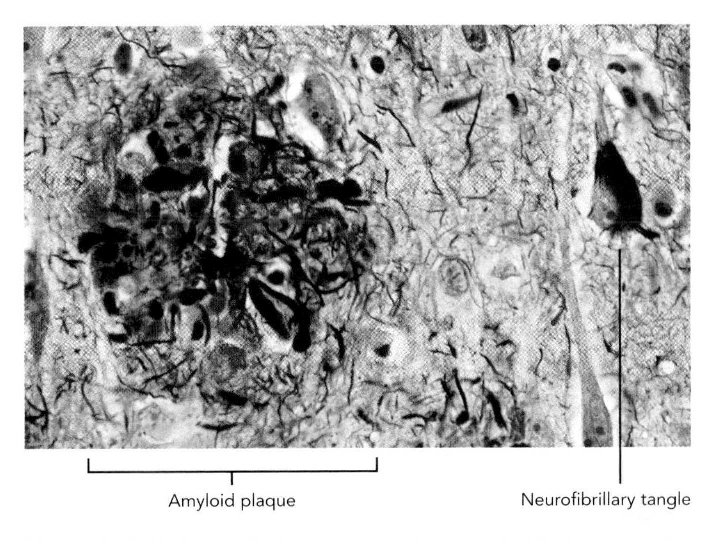

Figure 5.6. Enhanced photograph of an amyloid plaque and a neurofibrillary tangle in the brain

areas account for the cognitive decline and memory loss in people with Alzheimer's. At first, the brain is able to compensate well enough that even a family member can't tell the difference between someone who has this initial damage and someone who does not. Over time, however, as more and more connections are damaged and neurons begin to die, regions like the hippocampus disintegrate and the brain begins to lose crucial functions such as memory storage. Symptoms related to memory loss then become noticeable.

THE ROLE OF PROTEINS IN ALZHEIMER'S DISEASE

What causes plaques and tangles to form? Scientists have learned that the *amyloid-beta peptide* is responsible for forming amyloid plaques. This peptide is part of a much larger protein called the *amyloid precursor protein* (APP), which is thought to be lodged in the cell membrane of dendrites, the short, branching extensions of neurons (fig. 5.7). Two separate enzymes cut through the precursor protein, each in a different place, releasing the amyloid-beta peptide (fig. 5.7). Once released from the cell membrane, the peptide floats in the space outside the neuron.

It turns out that the production and liberation of the amyloid-beta

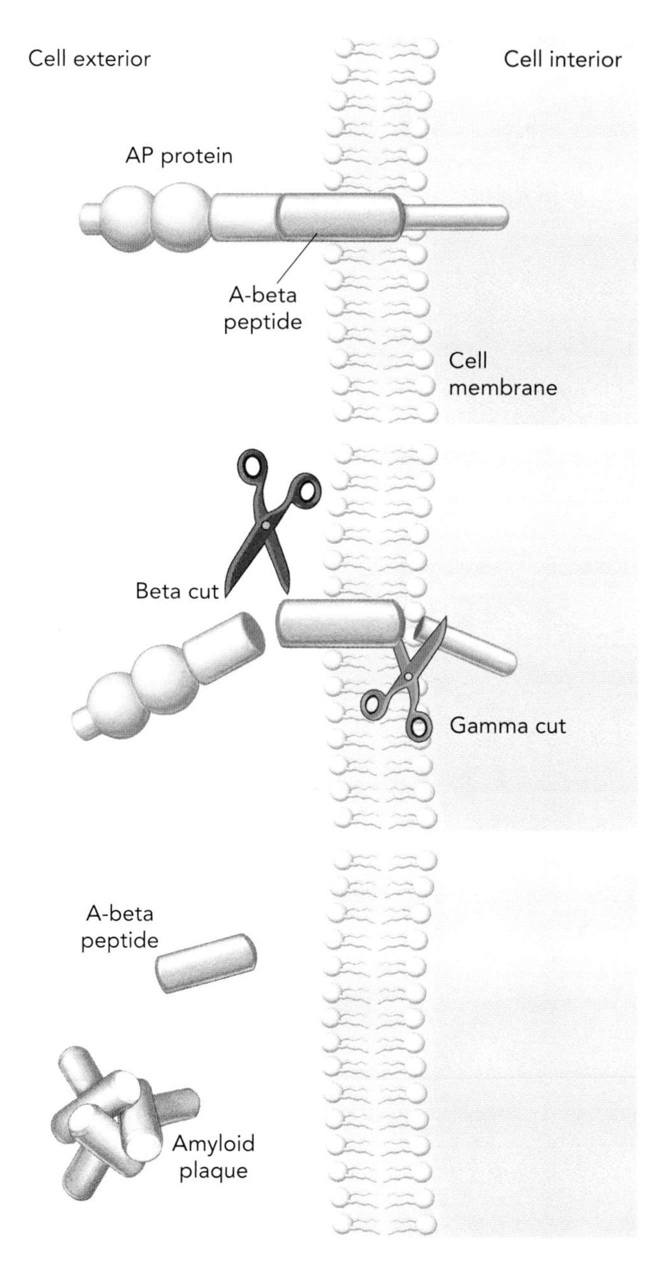

Figure 5.7. The amyloid precursor protein (APP), which is lodged in the cell membrane, contains the amyloid-beta (A-beta) peptide (*top*). Two enzymes make cuts through the amyloid precursor protein: the beta cut, followed by the gamma cut (*middle*). These cuts release the amyloid-beta peptide into the space outside the cell, where it may form amyloid plaques (*bottom*).

peptide are normal occurrences in everyone's brain. In people with Alzheimer's disease, however, production of the protein may be accelerated, or clearance of the protein from the area surrounding the cell may be slowed. Either action can result in abnormal accumulations of peptides. What's more, these peptides are sticky. They adhere to one another and ultimately form the amyloid plaques characteristic of Alzheimer's disease.

Another protein involved in Alzheimer's disease is called *tau*, and it is located inside the neuron. To function, a protein must have a three-dimensional shape. It assumes this shape by means of folding, a process in which the amino acids that make up the protein twist themselves into a very specific conformation. Think of it as exquisitely complicated origami. When a molecular defect causes the tau protein to misfold, it forms toxic clumps (fig. 5.8) that create neurofibrillary tangles.

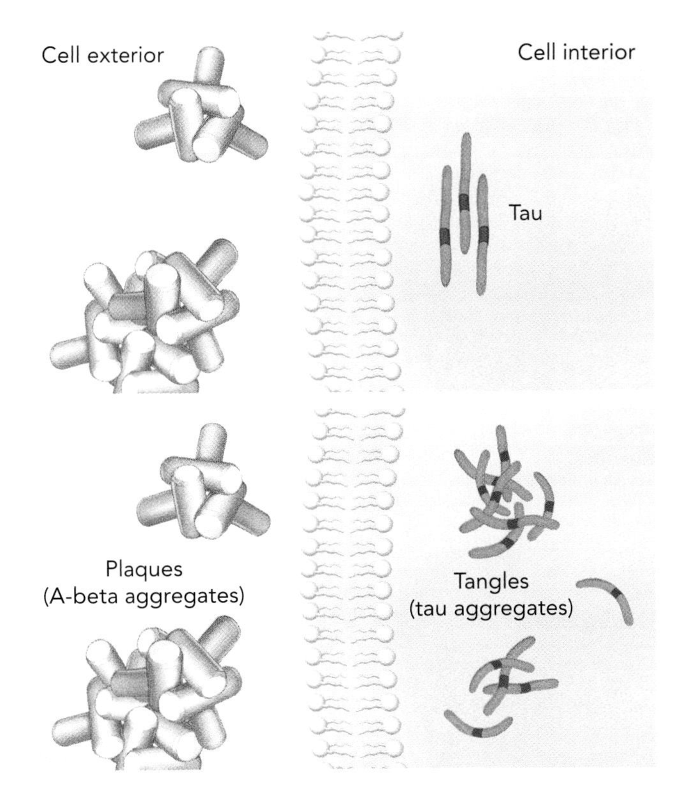

Figure 5.8. A molecular defect causes the tau protein to fold incorrectly. When this happens, the protein clumps inside the cell, forming neurofibrillary tangles.

The combination of these two types of aggregates—plaques outside the nerve cell and tangles within the nerve cell—causes the death of neurons and is responsible for the progression of Alzheimer's disease.

GENETIC STUDIES OF ALZHEIMER'S DISEASE

While Alzheimer's disease usually occurs in people in their seventies or eighties whose families have no history of the disease, a rare, early-onset form runs strongly in some families. John Hardy, now at University College London, had an unusual opportunity to study the genetic basis of Alzheimer's when Carol Jennings got in touch with him.

In the early 1980s Carol's father was diagnosed with Alzheimer's disease at age fifty-eight. Shortly thereafter, a sister and a brother of his, both in their midfifties, developed the disease. It turns out that Carol's great-grandfather had had the disease, as had her grandfather and a great-uncle. In the main branch of the family, five out of ten children had Alzheimer's disease, all at the same time. The average age of onset was about fifty-five (the record for early onset in familial Alzheimer's is the late twenties).

Hardy and his colleagues wanted to know what genes were inherited by all of the affected siblings in the Jennings family but not by any of the unaffected siblings. They found that the five affected siblings and an affected cousin shared an identical section of chromosome 21, the smallest chromosome in the human genome. But two of the unaffected siblings also had a little bit of that section of chromosome 21. This told Hardy that the gene responsible for Alzheimer's was *not* in the bit of chromosome 21 shared with the unaffected siblings. He then looked carefully at the part of chromosome 21 that had been inherited only by the family members with Alzheimer's, and there he found the defective gene that causes amyloid-beta peptides to clump.[11]

This was the first gene identified in Alzheimer's disease, and it opened up the study of the disease. Pathologists had already seen that amyloid-beta peptides form plaques, but Hardy showed that in the Jennings family the disease starts with a mutation in the gene for the amyloid precursor protein that causes the peptides to clump.

Hardy and other scientists have since found many additional mutations. A group of scientists in Toronto found families with inherited Alzheimer's who have mutations in the genes that code for a protein called presenilin.[12] These mutations prevent presenilin from helping to digest amyloid-beta peptides floating in the space between neurons. This finding fits together beautifully with Hardy's discovery. Both studies show that all of the families with early-onset Alzheimer's have mutations that lead to amyloid-beta peptides forming deadly clumps in the brain. Put another way, all of the mutations seem to converge on a single pathway that leads to early-onset, familial Alzheimer's (fig. 5.9).

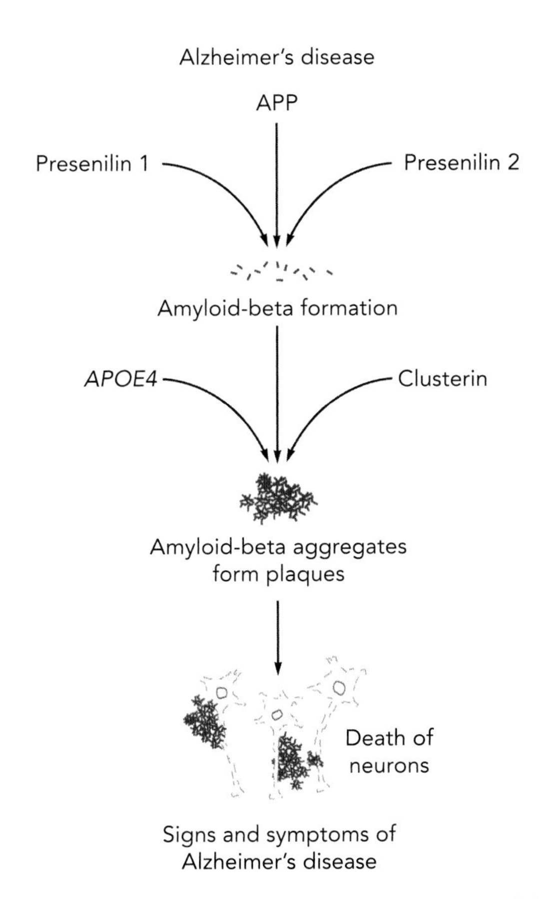

Figure 5.9. Several different pathways that lead to early-onset Alzheimer's disease converge to yield a common product: amyloid-beta aggregates. Clusterin is a type of protein that is produced in greater than usual amounts in people with Alzheimer's disease. It interacts with amyloid-beta peptides to exacerbate the loss of tissue in the entorhinal cortex.

These genetic studies of families with inherited Alzheimer's led scientists to wonder whether there might be mutations that reduce the number of amyloid-beta peptides. If such mutations exist, do they protect against Alzheimer's disease?

Thorlakur Jonsson and his colleagues at deCODE Genetics, a biotechnology company in Iceland, have found just such a mutation.[13] It causes one amino acid to be substituted for another in the amyloid precursor protein, and it results in fewer amyloid-beta peptides being generated. This mutation is particularly interesting because a different amino acid substitution at the same site on that precursor protein *causes* Alzheimer's disease. Even more fascinating, people over age eighty who have the protective mutation display better cognitive functioning than people of the same age who lack the mutation.

RISK FACTORS FOR ALZHEIMER'S DISEASE

Several scientists have been trying to work out the risk factors for the more common late-onset Alzheimer's. The most significant risk factor found to date is the *apolipoprotein E* (*APOE*) gene. This gene codes for a protein that combines with fats (lipids) to form a class of molecules called *lipoproteins*. Lipoproteins package cholesterol and other fats and carry them through the bloodstream. Normal amounts of cholesterol in the blood are essential for good health, but abnormal amounts can clog the arteries and give rise to strokes and heart attacks. One allele, or variation, of this gene is *APOE4*. The *APOE4* allele is rare in the general population, but it puts people at risk of developing late-onset Alzheimer's disease. In fact, about half of the people with late-onset Alzheimer's have this allele.

Since we can't change our genes, is there anything else we can do to lower the risk of developing Alzheimer's? One possibility has emerged recently, and it has to do with the way our body handles glucose as we age.

Glucose is the body's main source of energy, and it comes from the food we eat. The pancreas releases insulin, which essentially enables the muscles to absorb glucose. As we age, all of us become a bit insulin resistant, meaning that our muscles are less sensitive to the effects of insulin.

As a result, the pancreas tries to crank out a little bit more insulin, and this makes glucose regulation less stable. If glucose regulation becomes too unstable, we develop type 2 diabetes.

A number of studies have shown that type 2 diabetes is a risk factor for Alzheimer's disease. Furthermore, changes in glucose regulation that accompany type 2 diabetes seem to affect the areas of the hippocampus that are involved in age-related memory loss. The important thing is that we can actually modify these age-related changes through diet and physical exercise, which can increase our muscles' sensitivity to insulin and thus aid in the absorption of glucose.

Environmental factors and *comorbidities*, or other diseases that people have, may also contribute to the susceptibility to Alzheimer's disease, but all studies to date point to amyloid clumping as the fundamental cause of dementia. This is a very powerful hypothesis, and it has been extremely useful in guiding research. Recent studies have therefore focused on preventing clumping and clearing preexisting amyloid clumps by using antibodies that specifically recognize these clumps. As we have seen, disorders such as schizophrenia and depression seem to be caused not by a single gene but by hundreds of genes, so figuring out how those disorders come about is much more difficult. Even though it feels slow, our progress in understanding Alzheimer's disease has been amazingly rapid.

FRONTOTEMPORAL DEMENTIA

Alzheimer's disease is not the only common dementia. Another common form is frontotemporal dementia. Frontotemporal dementia was discovered a decade before Alzheimer's disease by Arnold Pick, a professor of psychiatry at the University of Prague. The disorder used to be considered rare, but we now know that it and Alzheimer's disease account for most cases of dementia in people over the age of sixty-four. Moreover, frontotemporal dementia is the most frequent cause of dementia in people *under* age sixty-five, affecting an estimated forty-five thousand to sixty-five thousand people in the United States. It generally begins at a younger age and progresses more rapidly than Alzheimer's disease.

Frontotemporal dementia begins in very small areas of the frontal lobe of the brain that are involved with social intelligence, particularly our ability to inhibit impulses (fig. 5.10). The disorder was once considered impossible to distinguish from Alzheimer's disease in a living person, but today that is no longer true. Frontotemporal dementia commonly results in profoundly disordered social behavior and moral reasoning. People may commit uncharacteristic antisocial acts, such as shoplifting. One study found that early in the illness, about half of all patients either were arrested or could have been arrested for something they had done. Such behavior is not characteristic of people with Alzheimer's disease.

Frontotemporal dementia also affects the parts of the brain that allow us to relate to others. People with this disorder who were once loving and kind may become indifferent to those around them. They also become vulnerable to addiction, engaging in regular overeating and taking up unhealthful habits, such as smoking. Sometimes they cannot control their spending and go bankrupt. This dementia has a huge impact on families because it affects people in midlife, many of whom have children.

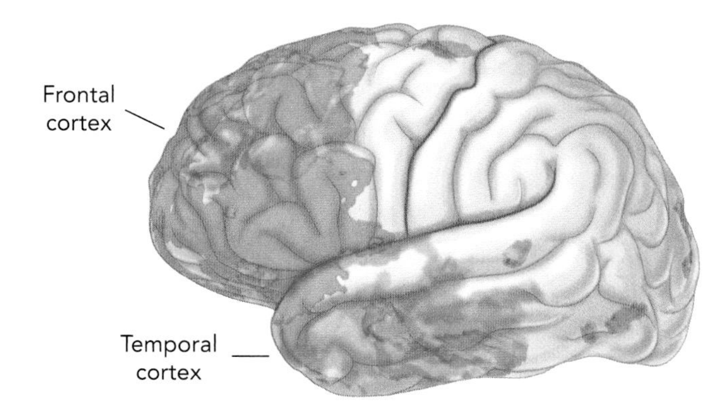

Figure 5.10. Frontotemporal dementia affects the frontal and temporal cortices of the brain.

THE GENETICS OF FRONTOTEMPORAL DEMENTIA

The biological mechanism of frontotemporal dementia—a disorder that results from damage to the frontal and temporal lobes—is the same as that of Alzheimer's disease: genetic mutations result in misfolded proteins that form clumps in the brain. That is why people with these two disorders have common symptoms. But some of the genes responsible for the protein misfolding are different in each disorder. The three mutated genes responsible for frontotemporal dementia are the gene that codes for the tau protein, the *C90RF72* gene, and the gene that codes for progranulin, a protein with several roles in the brain. Each mutated gene damages the same region of the brain, and each does it by means of abnormal protein folding (fig. 5.11).

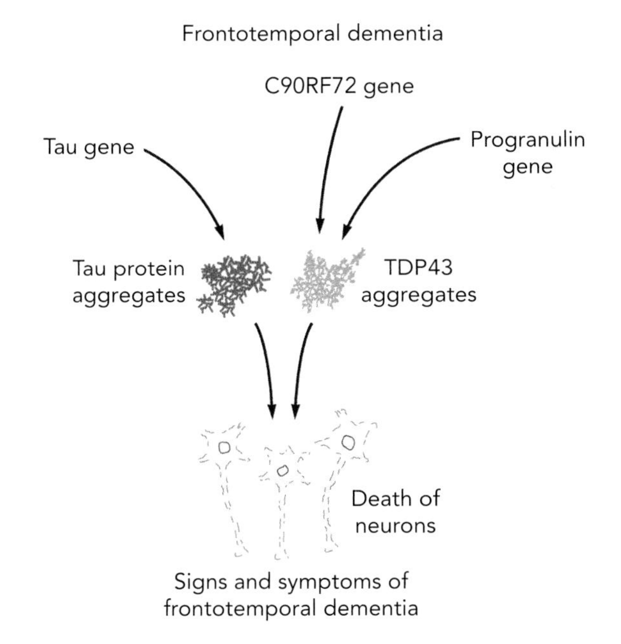

Figure 5.11. Mutations in three genes lead to frontotemporal dementia.

The mutated progranulin gene produces the normal progranulin protein, it just doesn't produce enough of it. (The normal progranulin protein is thought to keep another protein, TDP-43, from misfolding.) The simplicity of this mechanism is encouraging. It suggests that a plausible way to treat frontotemporal dementia is either to find a drug that will increase the amount of progranulin in the blood and brain or to figure out a way of delivering progranulin to the brain. In fact, Bruce Miller, of the University of California, San Francisco, who has studied frontotemporal dementia extensively, thinks it may be one of the simplest neurodegenerative diseases to treat. He is now testing drugs that are designed to elevate progranulin concentrations in the blood and brain.[14]

Miller has made an additional discovery, one that supports the findings of John Hughlings Jackson, a great nineteenth-century neurologist. Jackson was the first person to realize that the two hemispheres of the brain deal with different mental functions: the left hemisphere deals with logical functions such as language and numbers, and the right hemisphere deals with more-creative functions such as music and art. Moreover, Jackson suggested that the two hemispheres inhibit each other. Thus, damage to the left side of the brain would make it incapable of inhibiting the right side, thus freeing up the right side's creativity. Miller has described a series of patients whose frontotemporal dementia is restricted to the left hemisphere. A number of these people show outbursts of creativity, particularly people who were creatively inclined before the disease damaged their left hemisphere. The damage to the left hemisphere appears to have freed up the creative and musical right hemisphere.

These findings illustrate a remarkable principle of general brain function: when one neural circuit is turned off, another circuit may turn on. Why? Because the inactivated circuit normally inhibits the other circuit.

LOOKING AHEAD

The first scientist to describe a protein-folding disorder was Stanley Prusiner, who observed misfolding in the 1980s in Creutzfeldt-Jakob disease, a rare disorder. Other scientists, as we have seen, went on to show that protein misfolding contributes to Alzheimer's disease and frontotemporal dementia. At first glance, these dementias might seem to have little, if anything, in common with movement disorders. But a closer look reveals that Parkinson's disease and Huntington's disease also result from protein misfolding. We will turn to those brain disorders in chapter 7.

First, however, let us explore what brain disorders can tell us about another aspect of human nature: creativity. Just as our feelings, thoughts, behavior, social interactions, and memory have a biological basis, so, too, does our innate creativity. Earlier chapters touched on various expressions of creativity in people with autism, depression, bipolar disorder, and schizophrenia. Some people with Alzheimer's and frontotemporal dementia also express themselves creatively, most often in visual art. In chapter 6 we will explore what we have learned about creativity from artists with these brain disorders.

6

OUR INNATE CREATIVITY: BRAIN DISORDERS AND ART

Artists—painters, writers, sculptors, composers—seem different from other people, blessed with special gifts that the rest of us lack. The ancient Greeks believed that creative people were inspired by the muses, the goddesses of knowledge and the arts. The Romantic poets of the nineteenth century had a different view of creativity. They argued that creativity arises from mental illness, which diminishes the constraints posed by habit, convention, and rational thought and enables the artist to tap into unconscious creative powers.

Today, we know that creativity originates in the brain. It has a biological basis. We also know that while certain forms of creativity arise in association with mental disorders, our creative capability is not dependent upon mental disorder. Moreover, the capability for creativity is universal. Each of us, in various ways and with varying degrees of skill, expresses it.

Yet the Romantics weren't entirely wrong. For most people, our innate creative capability is not easily summoned. Scientists have yet to uncover the biological mechanisms of creativity, but they have discovered some of its precursors, one of which seems to be divesting ourselves of inhibitions, allowing our minds to wander more freely and to seek new connections between ideas. Such communion with the unconscious is shared by all creative people, but it is sometimes particularly striking in creative people with mental disorders.

This chapter explores what brain disorders, both psychiatric and neurological, can tell us about our creative capability. We begin by examining creativity from several perspectives. First, we focus on the work of an extraordinarily gifted contemporary artist. Next, we approach creativity from the perspective of the viewer. Finally, we explore what we have learned about the nature of the creative process and the biology of creativity.

In earlier chapters we saw people with schizophrenia, depression, and bipolar disorder who express their creative gifts in art, literature, and science. This chapter focuses primarily on the visual art of patients with schizophrenia—so-called psychotic art—not only because it is beautiful and moving but because it has been collected and studied extensively. We go on to explore the influence of this art on modern art, notably on Dada and Surrealism. We then touch on the creativity of people with other brain disorders: bipolar disorder, autism, Alzheimer's disease, and frontotemporal dementia. We conclude with some initial insights into what modern brain studies have shown about our innate creative capability.

PERSPECTIVES ON CREATIVITY

THE ARTIST

Chuck Close is dyslexic, and as a child there were many things he felt he couldn't do. One thing he could do, however—and do well—was draw. He became particularly interested in drawing faces, which is intriguing because Close is also face-blind—that is, he can recognize a face as a face, but he cannot associate that face with a particular person.

Our ability to recognize faces resides in the right fusiform gyrus of the inferior medial temporal lobe of the brain. People with damage to the front of that region are face-blind, like Close. People with damage to the back of that region cannot see a face at all. Close is probably the only person in the history of Western art to paint portraits without being able to recognize individual people. Why, then, did he focus on being a portrait artist? Close says his art was an attempt to make sense of a world he didn't understand. For him, it's not so strange that he makes portraits. He was driven to make portraits because he was trying to understand the faces of people he knows and loves and commit them to memory. For him a face has to be flattened out. Once he flattens it, he can commit it to memory in a way that he cannot if he's looking at it head-on. If he looks at you and you move your head half an inch, it's a new head for him that he has never seen before. But if he takes a photograph of the face and flattens it out, he now can effect the translation from one flat medium to another.

That translation goes like this. First, Close takes a photograph of a face. Then he covers the photograph with a sheet of transparent Plexiglas and pixelates it—that is, he applies a grid that separates the image into thousands of tiny cells. Finally, he paints each of the tiny pixels, row

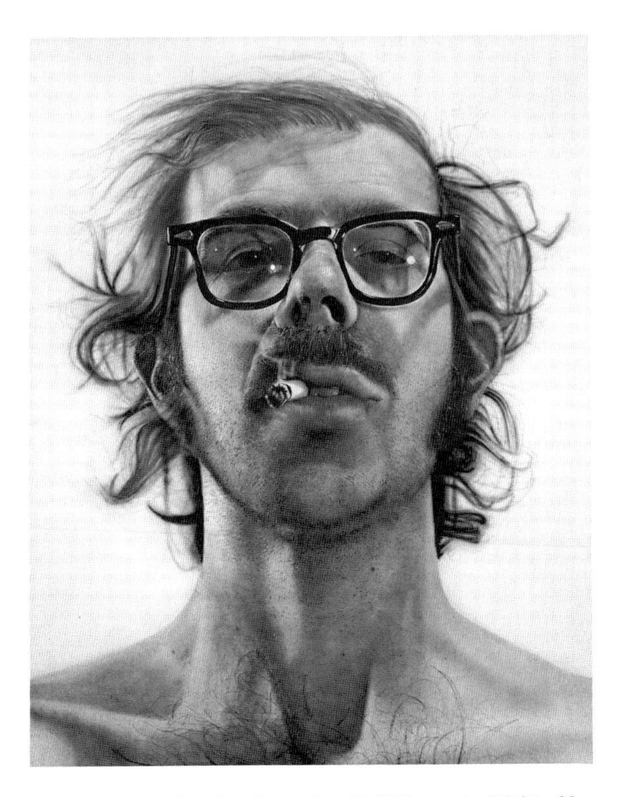

Figure 6.1. Chuck Close, *Big Self-Portrait*, 1967–68. Acrylic on canvas, 107.5 × 83.5 × 2 inches

by row, which together coalesce into a portrait. The final image is clearly composed of its constituent parts.

In his early work, Close used this method to achieve an unprecedented level of realism (fig. 6.1), consistent with his desire to make sense of the world. In time, however, he began using the grid more experimentally, revealing a progressive lifting of inhibitions. First, he filled each cell with a repeated mark, a dot, to create wonderfully complex portraits from very simple, incremental units. Eventually, the technique evolved into painting each cell as a tiny abstract painting made of concentric circles (fig. 6.2). Rather than coloring each square with the same uniform flesh tone, Close made several saturated rings; from a distance, they create the illusion of a single color and create a vibrant, believable portrait.

Studies have shown that the right hemisphere of the brain is more concerned with putting ideas together, seeing new combinations—in short, with aspects of creativity. The left hemisphere is concerned with language and logic. As we saw in chapter 5, John Hughlings Jackson, the founder of modern neurology, argued a century ago that the left hemisphere of the brain inhibits the right hemisphere, and as a result, damage to the left hemisphere can enhance creativity. Close's left hemisphere is compromised, as is evident in his dyslexia, and like many other artists, he is left-handed, which further indicates that the right hemisphere of his brain is dominant.

Not only has Close taken full advantage of this possible pathway to creativity, he has worked, as a gifted athlete might, at doing better what he already does well. He has used his dyslexia to enhance his artistic strengths. He has pointed out that everything he does is driven by his learning disabilities. He didn't take algebra, geometry, physics, or chemistry. He got through life by taking extra-credit art courses and projects to show his teacher that he was interested in his classwork, even though he couldn't recall facts later on. He got reinforcement for demonstrating that he had skills, and this made him feel special. As a result, his artistic ability is extraordinary, and his depiction of faces is continually evolving.

Close exemplifies two important aspects of creativity in addition to the lifting of inhibitions: the determination to work hard and overcome difficulty, and the enormous plasticity of our brain. As we saw in the

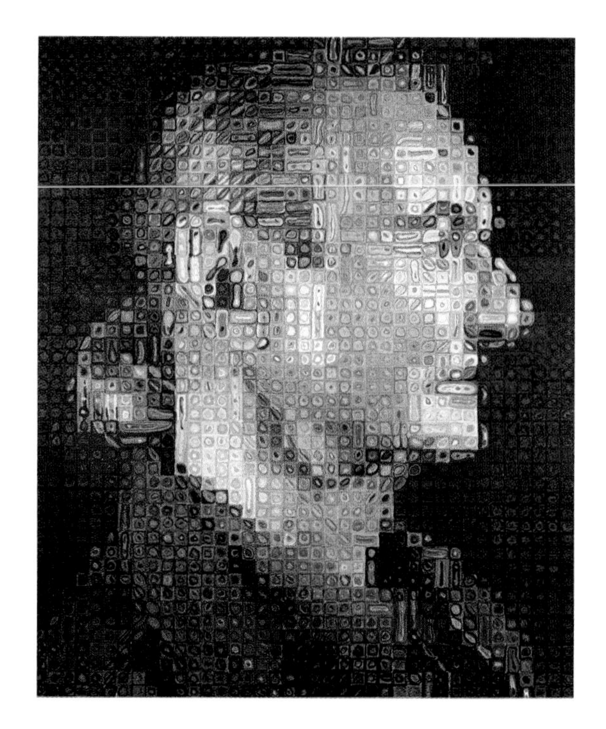

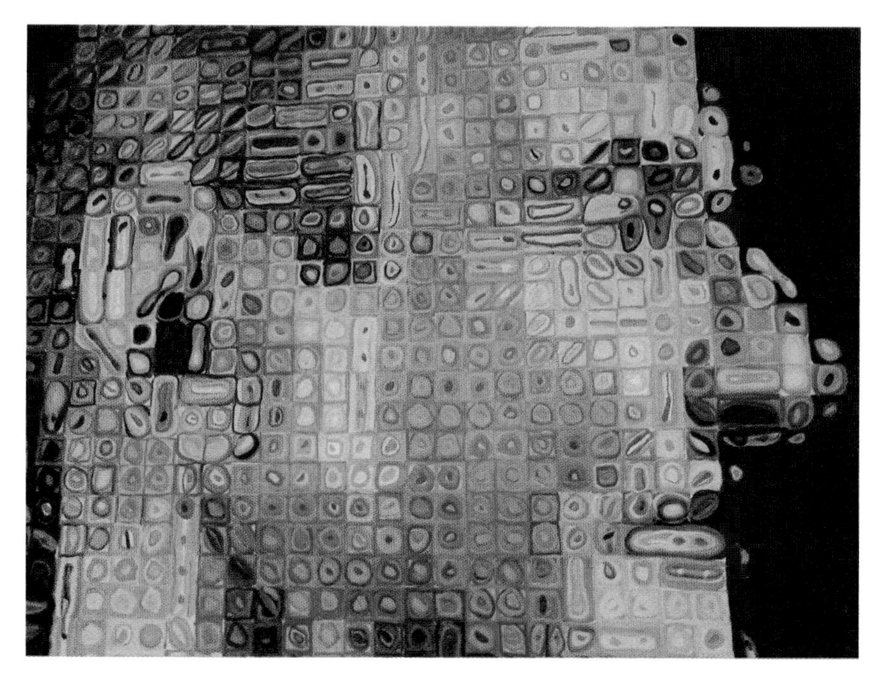

Figure 6.2. Chuck Close, *Roy II*, 1994. Oil on canvas, 102 × 84 inches (*top*), and a detail from the same painting (*bottom*)

chapters on autism and Alzheimer's disease, damage to some regions of the brain can be compensated for by increased strength and effectiveness in other regions. The brain's ability to compensate for damage can also enhance an artist's ability to do new, more interesting and creative things.

THE VIEWER

While the ancient Greeks and the Romantics were fascinated by the creative artist, it wasn't until the turn of the twentieth century that the viewer's experience of art took center stage. The idea that both the beholder and the artist engage in creative mental processes was first introduced around 1900 by Alois Riegl, a founder of the Vienna School of Art History.

Riegl and his two great disciples, Ernst Kris, who later became a psychoanalyst, and Ernst Gombrich, argued that when we look at a work of art, each of us sees it in slightly different terms. That is because there is ambiguity in almost every object we see, but particularly in great works of art. Each of us interprets that ambiguity differently, and as a result, each of us sees a given work of art differently. This implies that we each create our own view of the work—that is, we undergo a creative process that is similar in nature, although more modest in scope, to the creative process of the artist. This creative process is known as the *beholder's share.*

We know this is true because, as we have seen, the actual sensory information going from any image to our brain is rudimentary, fragmentary. Our eyes are not a camera that relays a complete image to our brain. Rather, our brain receives incomplete sensory information and interprets it in light of our emotions, experience, and memory. This interpretive process, carried out by our brain, is what enables us to reconstruct our own unique perception of the image we see, and it is the basis of the beholder's share.

Ann Temkin, chief curator of painting and sculpture at the Museum of Modern Art in New York, uses Close's portrait of Roy Lichtenstein (fig. 6.2) as an example of the viewer's response. "There is obviously a back-and-forth in these pictures between the abstract marks, the painting act, and the representation of somebody," she says. "Neither one is the whole experience. Part of the experience is the abstract circles and squares and funny shapes that you see close up, and part of it is stepping back and recognizing that Oh, it's Lichtenstein. The process that you go through in recognizing Lichtenstein is so embedded in the painting that as the viewer you are almost made to re-create it."[1] That process of recognition is also embedded in the way our brain constructs Lichtenstein's face from Close's tiny geometric shapes.

THE CREATIVE PROCESS

Is there an explanation for why a burst of creativity occurs at certain times in history and in certain places? Whether we are talking about the cultural ferment of the Renaissance, the Impressionists in Paris, the Figurative Expressionists of Vienna 1900, or the Abstract Expressionists in New York, interaction among creative people is essential. Sometimes that interaction comes in the form of a rivalry among colleagues, or conversely the desire to support one another. Ideas commonly emerge when creative people talk with one another in a café or at a party. In other words, the myth of the isolated genius is just that: a myth.

What, then, are the factors that contribute to individual creativity? For Close, as we have seen, the essential aspect of creativity is problem solving: technical competence and a willingness to work hard. Studies have found that certain additional features increase the likelihood of creativity. The first is personality: some personality types are more likely to be creative than others. Note the plural—creativity is not limited to a single personality type, as the developmental psychologist Howard Gardner emphasizes in his work on multiple intelligences. Rather, creativity comes in multiple forms: some of us are strong in arithmetic skills, others in language skills, others in visual skills.[2]

The second feature is the period of preparation, when a person works on a problem both consciously and unconsciously. The third feature is the initial moment of creativity, the Aha! moment, when a sudden insight connects previously unassociated factors in a person's brain. Last is the subsequent working through of the idea.

After consciously working on a problem, we need an incubation period, when we refrain from conscious thought and let our unconscious roam. This incubation period, says the psychologist Jonathan Schooler, is for "letting the mind wander."[3] New ideas often come to us not when we are hard at work on a project but when we are going for a walk, taking a shower, thinking about something else. These are the Aha! moments, the epiphanies, of creativity, and we are now beginning to get some insight into the biology underlying them.

Kris, a student of unconscious mental processes in creativity, observed that creative people experience moments in their work in which they undergo, in a controlled manner, a relatively free communication between the unconscious and conscious parts of their mind. He calls this controlled access to our unconscious "regression in the service of the ego."[4] It means that creative people go back to a more primitive form of psychological functioning, one that allows them access to their unconscious drives and desires—and to some of the creative potential associated with them. Because unconscious thinking is freer and more likely to be associative—it is characterized by images as opposed to abstract concepts—it facilitates the emergence of Aha! moments that promote new combinations and permutations of ideas.

THE BIOLOGY OF CREATIVITY

Although we know little about the biology of creativity, it is clear that creativity entails the lifting of inhibitions. Jackson's idea that the left and right hemispheres of the brain inhibit each other and that damage to the left hemisphere frees up the creative capabilities of the right hemisphere has been validated by modern technology.

PET scans of the brain, for example, have revealed a fascinating difference in the way the left and right hemispheres respond to a repeated stimulus. The left hemisphere always responds to the stimulus (a word or an object), regardless of how often it is presented. The right hemisphere, in contrast, is often bored with the routine stimulus but responds actively to novel stimuli. Thus, the right hemisphere, which is more concerned with novelty, has a greater capability for creativity. Similarly, the neurologist Bruce Miller, whom we encountered in chapter 5, made the remarkable discovery that people with frontotemporal dementia in the left hemisphere sometimes undergo a burst of creativity, presumably because the disorder in the left hemisphere is removing its inhibitory constraint over the right hemisphere.[5]

This idea has been carried further in a very interesting collaboration between Mark Jung-Beeman at Northwestern University and John Kounios at Drexel University. They presented study participants with problems that could be solved either systematically or by means of a sudden Aha! insight. When participants call on an Aha! insight, a region of their right hemisphere lights up. These experiments, although in the early stages, support the idea that sudden flashes of insight, moments of creativity, occur when our brain engages distinct neural and cognitive processes, some of which are located in the right hemisphere.[6]

A similar lesson emerges from brain-imaging experiments carried out by Charles Limb and Allen Braun at the National Institutes of Health. They wanted to understand the differences in the mental processes underlying jazz improvisation, on the one hand, and the performance of a memorized musical sequence, on the other. They put experienced jazz pianists into a scanner and asked them to play a musical sequence they had created on the spot or a tune they had memorized. Limb and Braun found that improvisation relies on a characteristic set of changes in the dorsolateral prefrontal cortex, an area concerned with impulse control.[7]

How does impulse relate to creativity? Limb and Braun found that before the pianists began to improvise, their brain showed a "deactivation" of the dorsolateral prefrontal cortex. However, when they were playing the memorized tune, this region remained active. In other words, while they were improvising, their brain was damping down their inhibitions normally mediated by the dorsolateral prefrontal cortex. They were able to create new music in part because they were uninhibited and not self-conscious about being creative.

Simply turning off the dorsolateral prefrontal cortex won't turn any of us into a great pianist, however. These pianists benefited from the lifting of inhibitions only because they, like most other successful creative people, had spent years practicing their art form, filling their brains with musical ideas that they could spontaneously recombine onstage.

THE ART OF PEOPLE WITH SCHIZOPHRENIA

The Romantic movement, which blossomed in the first half of the nineteenth century, emphasized intuition and emotion over rationalism as a source of aesthetic experience and awakened a keen interest in the creativity of people with mental illness. Romanticism characterized psychoses as exalted states that free a person from conventional reason and social mores and provide access to hidden realms of the mind that are normally unconscious and thus inaccessible.

The first person to take an interest in the art of psychotic patients was actually Philippe Pinel, the physician who developed a humane, psychological approach to mental patients. In 1801 he wrote about the art of two of his psychiatric patients and concluded that insanity can sometimes unearth hidden artistic talents.[8] In 1812 Benjamin Rush, a Founding Father of the United States and the founder of psychiatry as a distinct discipline in the United States, echoed Pinel's view. Insanity, Rush wrote, is like an earthquake that "by convulsing the upper strata of our globe, throws upon its surface precious and splendid fossils, the existence of which was unknown to the proprietors of the soil in which they were buried."[9]

In 1864 the Italian physician and criminologist Cesare Lombroso collected works of art from 108 patients and published *Genio e Follia*, or "Genius and Madness," later translated into English as *The Man of Genius*. Like Rush, Lombroso found that insanity transformed some people who had never painted before into painters, but Lombroso saw this art as part of the patient's illness and was insensitive to its aesthetic merits.[10]

Emil Kraepelin, the father of modern scientific psychiatry, took a less Romantic, though no less appreciative, approach to the relationship between psychosis and creativity. Soon after becoming director of the psychiatric clinic of the University of Heidelberg, in 1891, Kraepelin noted that some of his schizophrenic patients painted. He started to collect the art of these patients as a *Lehrsammlung*, a teaching collection, to see whether studying the paintings might aid physicians in diagnosing the disorder. Kraepelin also thought that painting might be therapeutic for patients, a view that now has considerable support.

Karl Wilmanns, a subsequent director of the Heidelberg clinic, continued Kraepelin's tradition of collecting the paintings of his psychotic patients and in 1919 recruited Hans Prinzhorn to work on the collection. Prinzhorn was a psychiatrist and art historian who had trained in art history under Alois Riegl.

Prinzhorn proceeded to expand the collection. Since only about 2 percent of the inmates at the Heidelberg clinic were creating art, he asked directors of other psychiatric institutions—in Germany, Austria, Switzerland, Italy, and the Netherlands—to send him the artwork of their psychotic patients. As a result of this appeal, Prinzhorn received more than five thousand paintings, drawings, sculptures, and collages representing the work of about five hundred patients.

The patients whose art Prinzhorn collected had two salient characteristics: they were psychotic, and they were artistically naïve, that is, untrained in art. Prinzhorn recognized that the art of psychotic patients is not simply pathology translated into a visual language. The lack of artistic training evidenced in most of their drawings is no different from what we would see in the work of any inexperienced adult who took up drawing; it reflects nothing pathological in and of itself. Prinzhorn realized that the patients' images were creative works in their own right and that they were remarkable examples of naïve art.

As Prinzhorn was careful to point out, however, artistic naïveté is not confined to artists who suffer from psychosis. One of the most notable examples of an untrained artist who was not psychotic is Henri Rousseau (1844–1910). Rousseau, a French toll collector, was often ridiculed by critics during his lifetime, but his work is of extraordinary artistic quality. He eventually came to be recognized as a self-taught genius and major Post-Impressionist painter (figs. 6.3 and 6.4), and his work influenced several generations of artists, including the Surrealists and Picasso. Although Rousseau never actually left France, his best-known paintings depict jungle scenes (fig. 6.4). He drew inspiration for these scenes from his unconscious fantasy life.

In the early twentieth century, psychiatric patients who were hospitalized commonly spent the rest of their lives—twenty to forty years—in an institution. Some of them began to paint after they were hospitalized. Rudolf Arnheim, a distinguished student of the psychology of art, notes:

> [T]housands of institutionalized patients seized on pieces of stationery or toilet paper, wrappers, bread, or wood to give visible

Figure 6.3. Henri Rousseau, *The Sleeping Gypsy*, 1897

Figure 6.4. Henri Rousseau, *The Flamingoes*, 1907

> expression to the powerful feelings of mental upheaval generated by their anguish, their frustration, their protests against confinement, and their megalomaniac visions. Yet among psychiatrists only an occasional prophetic forerunner sensed the diagnostic possibilities of those uncanny images and perhaps speculated on their oblique significance for the nature of human creativity.[11]

Prinzhorn's appreciation of the creativity and aesthetic value of his patients' art established that many aspects of what was then termed "psychotic art" are not mere curiosities but worthy of serious study. As Thomas Roeske, the current director of the Prinzhorn Collection, points out, the paintings gave a voice to people who otherwise would not have been heard from—and their voice was often quite distinctive.[12]

PRINZHORN'S SCHIZOPHRENIC MASTERS

In 1922 Prinzhorn published his highly influential book *Artistry of the Mentally Ill: A Contribution to the Psychology and Psychopathology of Configuration*, which he illustrated with examples from the Heidelberg collection.[13] Of the five hundred artists represented in the collection, 70 percent had schizophrenia; the remaining 30 percent had bipolar disorder. These proportions reflect, in part, the hospitalization rates of people with those psychiatric illnesses. Prinzhorn focused in particular on the work of ten patients whom he referred to as "schizophrenic masters." He presented the clinical history of each artist, protected by a pseudonym, followed by an analysis of the artist's work and of its clinical implications for diagnosis and for the course of the artist's disease.

Prinzhorn describes these patients as suffering from "complete autistic isolation . . . the essence of schizophrenic configuration,"[14] and he found their work to be characterized by a "disquieting feeling of strangeness."[15] For Prinzhorn, their art reflected the "eruptions of a universal human creative urge"[16] that counteracted the sense of isolation they were experiencing. Because most of his artists were untrained, Prinzhorn also used their art to demonstrate surprising parallels with the work of children and with the work of artists from primitive societies. In every case, the artworks reflect the unschooled artistic creativity present in all of us. For these artists, a blank piece of paper often represented a passive emptiness that cried out to be filled. As a result, they tended to cover every inch of the surface. We see this in paintings by three of Prinzhorn's schizophrenic masters, Peter Moog (fig. 6.5), Viktor Orth (fig. 6.6), and August Natterer (fig. 6.7).

Moog was born in 1871 and grew up in poverty. His father was thought to be mentally disturbed, but Moog himself was kind and very bright, with a strong memory. After leaving school, he became a waiter and began to lead a loose life filled with wine, women, and song. During this time he contracted gonorrhea. He married in 1900, but his wife died in 1907. While working as a manager of a large hotel, he started to drink heavily, and in 1908 he suddenly experienced a psychotic episode. A few weeks later he was diagnosed with schizophrenia and committed

Figure 6.5. Peter Moog, *Altar with Priest and Madonna*

to an asylum, where he lived until his death in 1930. Moog's vision, as we see in *Altar with Priest and Madonna* (fig. 6.5), is dominated by religious imagery.[17]

Orth was born in 1853 to an ancient, noble family. He developed normally as a child and later became a naval cadet, but he began to be plagued with paranoia when he was twenty-five years old and was hospitalized from 1883 until his death in 1919. At various times he believed himself to be the King of Saxony, the King of Poland, and the Duke of Luxembourg. In 1900 he began to paint. As Prinzhorn said of Orth's paintings, in his eagerness "no empty surface is safe." He covered every inch of the page, much as Moog did, but unlike Moog he did not create intricately detailed drawings. Many of Orth's paintings were seascapes and featured a three-masted ship that Prinzhorn believed was his training ship. In figure 6.6 we see an abstract version of a three-masted ship on the sea. The colorful diagonal areas "together give the effect of a mild sunset at sea," wrote Prinzhorn.[18]

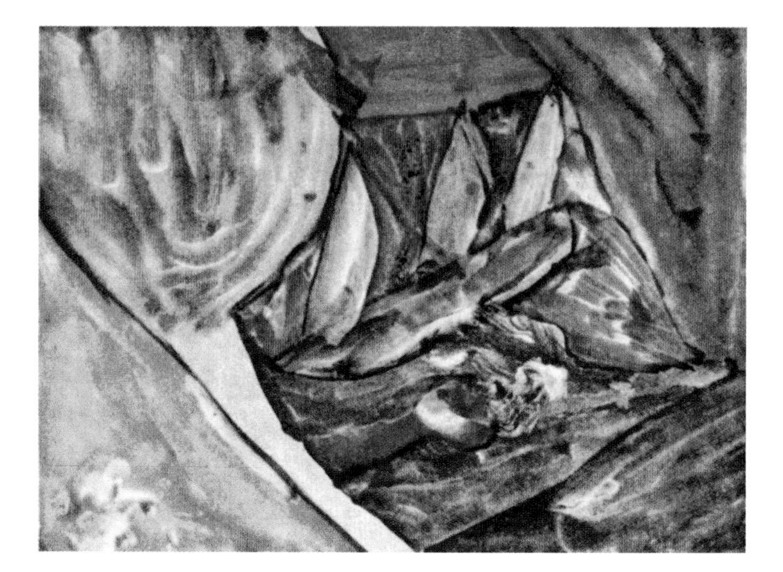

Figure 6.6. Viktor Orth, *Barque Evening at Sea*, watercolor, 29×21 centimeters

Another of Prinzhorn's schizophrenic masters was August Neter, born August Natterer in 1868 in Germany. He studied engineering, married, and was a successful electrician, but he suddenly developed anxiety attacks accompanied by delusions. On April 1, 1907, he had a critical hallucination of the Last Judgment, during which he said ten thousand pictures flashed by in half an hour. "The pictures were manifestations of the last judgment," said Natterer. "They were revealed to me by God for the completion of [Christ's] redemption."[19]

Natterer attempted to capture in his artwork the ten thousand images of his Last Judgment hallucination. The images are always executed in a clear, objective style almost like a technical drawing, as they are in *Axle of the World with Rabbit* (fig. 6.7). Natterer insisted that this painting predicted the First World War—he knew everything in advance. According to Natterer, the rabbit in the painting represents "the uncertainty of good fortune. It began to run on the roller . . . the rabbit was then changed into a zebra (upper part striped) and then into a donkey (donkey's head) made of glass. A napkin was hung on the donkey; it was shaved."[20]

Figure 6.7. August Natterer, *Axle of the World with Rabbit*, 1919

SOME CHARACTERISTICS OF PSYCHOTIC ART

The artworks pictured on the preceding pages very likely spring from the same type of intrinsic creative capability as any other work of art, but because the artists suffered from schizophrenia and were unfettered by artistic or social conventions, their work was thought by critics of the time to be a purer expression of their unconscious conflicts and desires. That is why most people have such a powerful emotional response to their art. That is also why these works strike us, even with our modern sensibility, as amazingly original. In fact, the publication of these works in the early 1920s caused people to reconsider the idea of "original" in Western art. Much of what we consider art, Roeske argues, is ideologically charged: "We expect certain things from art." He goes on to say, "The Prinzhorn collection delivers many more aspects of the life of the individual and of life in society than conventional art can do."[21]

What makes the art that Prinzhorn collected different from that of other artists, trained or untrained? Schizophrenia, as we know, results in disordered thought, which detaches an individual from reality. This disturbance in the relation between a person and his or her social environment can lead to striking distortions of outlook, distortions that frequently alter the function of artistic form. Thus, one common characteristic of schizophrenic art is the juxtaposition of unrelated elements. Another is the depiction of delusions and hallucinatory images. Still others are ambiguous images or the reassembling of dismembered body parts. The work of each artist is characterized by recurring motifs that spring from his or her unconscious mind. Thus the works do have, as Kraepelin had predicted, distinctive themes that are specific to their creators.

THE IMPACT OF PSYCHOTIC ART ON MODERN ART

The Dada movement and the subsequent Surrealist movement emerged largely in response to the carnage of World War I. It is difficult to overestimate the psychic effects of the Great War. When the war began, many young people entered it enthusiastically, believing that war would lead to the rejuvenation of society. But within a year many people were left with a sense of utter, senseless destruction. The war called into question the belief in the inevitability of social progress; even more important, it struck at the heart of Western rational self-understanding. From the failure of reason there emerged the possibility that irrationality might be a life-affirming alternative.

It was amid the chaos of the war that Dada emerged in Zurich in 1916. Surrealism originated shortly thereafter in Paris, where most adherents of Dada settled after the war. Although it was first conceived of as a literary movement, the techniques and orientation of Surrealism proved better suited to art. Like the Dadaists, the Surrealists were opposed to the tradition of academic art and the values for which it stood, but they were in search of a new, more creative and positive philosophy than the chaos of Dada. They found such a philosophy in the work of Freud, Prinzhorn, and similar thinkers.

Freud had documented the importance of unconscious thought, which is not rational and is not governed by a sense of time, space, or logic. Moreover, he pointed to dreams as the royal road to the unconscious. The Surrealists attempted to eliminate logic from their work and to draw on dreams and myths for inspiration, thus unlocking the power of the imagination. In addition, they were determined, as were Cézanne and the Cubists after him, to move art from its historical representational trajectory onto a new one.

Max Ernst, a leader first of Dada and later of Surrealist art, bought a copy of Prinzhorn's book and took it to Paris, where it became the "Picture Bible" of the Surrealists. Although most of the Parisian members of the Surrealist group could not read German, the images in Prinzhorn's book spoke for themselves, illustrating what could be accomplished outside conventional bourgeois attitudes and inhibitions.

The complete naïveté of the psychotic artists was a powerful stimulus to the Surrealists. They set out to liberate creativity from the limitations of rational thought by exploring the hidden depths of the unconscious mind. They encouraged one another to explore and to express their own erotic and aggressive drives. Accordingly, every Surrealist artist relied on central motifs that derived from his or her distinctive, unconscious mental processes, just as the psychotic artists did.

In 2009 Roeske put together an exhibition in Heidelberg in which he systematically compared Surrealist art and psychotic art from the Prinzhorn collection. The exhibition, "Surrealism and Madness," focused on four processes, or techniques, that the Surrealists employed to tap into the unconscious, thereby emulating psychotic artists.

Figure 6.8. Heinrich Hermann Mebes, *Das brütende Rebhühn oder die herrschende Sünde* (*top*); Frida Kahlo, *Without Hope*, 1945 (*bottom*)

The first and most important process was *automatic drawing*. Automatism is a method of tapping into the unconscious that was introduced by psychiatrists in the nineteenth century. André Masson was a pioneer in automatic drawing. The second process was *combining unrelated elements*. The more distant the relationship between the elements, the truer and stronger the image. Ernst took the technique to an astonishing level of virtuosity in his Dada collages. Roeske compared an image by Heinrich Hermann Mebes from the Prinzhorn collection to a work by Frida Kahlo (fig. 6.8).

The third process, known as the *paranoiac-critical method*, was developed by Salvador Dalí. Dalí attributes the visual double meaning in

Figure 6.9. August Natterer, *Axle of the World with Rabbit*, 1919 (*top*); Salvador Dalí, *Remorse, or Sphinx Embedded in the Sand*, 1931 (*bottom*)

his paintings, which are essentially picture puzzles, to the change in perception brought about by paranoia. Similar ambiguities can be found in pictures from the Prinzhorn collection. In the exhibition, Roeske placed a work by Dalí next to Natterer's *Axle of the World with Rabbit* (fig. 6.9).

The fourth process was *figure amalgamation*, in which dismembered body parts are rearranged and fused, often to shocking effect. The Surrealist Hans Bellmer used this technique in his drawings.

The Surrealists aimed to create a pictorial art that already existed in the art of psychotic patients by devising ways of tapping into their own unconscious mind. Whereas the psychotic artists did this naturally and unselfconsciously, the Surrealists' deliberate efforts also succeeded, as Roeske's exhibition demonstrates. Both groups of artists evoke in us the "disquieting feeling of strangeness" that Prinzhorn described. Moreover, whereas the psychotic artists were untrained, the Surrealists went to great lengths to unlearn their training. Picasso claimed that he used to draw like Raphael and it took him a full lifetime to learn to draw like a child.[22]

WHAT OTHER BRAIN DISORDERS TELL US ABOUT CREATIVITY

The idea that creativity derives from madness has been nourished for centuries by the unusual prevalence of mood disorders among writers and artists. A different sort of genius, the savant, has been observed among people on the autism spectrum. Even neurological disorders, such as Alzheimer's disease and frontotemporal dementia, can uncover creative capability.

In her book *Touched with Fire: Manic-Depressive Illness and the Artistic Temperament*, Kay Redfield Jamison reviews the extensive body of research suggesting that writers and artists show a vastly higher rate of manic-depressive, or bipolar, disorder than the general population.[23] For example, Vincent van Gogh and Edvard Munch, two of the founders of Expressionism, both suffered from manic-depressive illness, as did the Romantic poet Lord Byron and the novelist Virginia Woolf. Nancy Andreasen, a psychiatrist at the University of Iowa, has examined creativity in living writers and found that they are four times as likely to have bipolar disorder and three times as likely to have depression as people who are not creative.[24]

Jamison points out that people with bipolar disorder do not have symptoms much of the time, but as they swing from depression to mania, they experience an exhilarating feeling of energy and a capability for formulating ideas that dramatically enhance their artistic creativity. The tension and transition between changing mood states, as well as the sustenance and discipline that people with bipolar disorder draw from periods of health, are critically important. Some people have argued that it is these tensions and transitions that ultimately give an artist with bipolar disorder his or her creative power.[25]

Ruth Richards at Harvard University has carried the analysis further.[26] She tested the idea that a genetic vulnerability to bipolar disorder might be accompanied by a predisposition to creativity. She examined patients' first-degree relatives who did not have bipolar disorder and found that the correlation indeed exists. Richards therefore proposes that genes which confer a greater risk of bipolar disorder may also confer a greater likelihood of creativity. This is not to imply that bipolar disorder creates the predisposition to creativity, but rather that people who have the genes associated with bipolar disorder also have the heightened exuberance, enthusiasm, and energy that express themselves in and contribute to creativity. These studies emphasize the importance of genetic factors in contributing to creativity.

CREATIVITY IN PEOPLE WITH AUTISM

People on the autism spectrum approach creative problem solving differently than neurotypical people do. In a study of both neurotypical people and people on the autism spectrum, Martin Doherty of the University of East Anglia in England and his colleagues found that people with a high number of autism traits generate fewer, but more original, ideas. He proposes that they are more likely to go directly to less common ideas because they rely less on associations or memory, which would constrain their creative thinking.[27]

In one test, study participants were asked to identify as many potential uses for a paper clip as they could. Many people said that a paper clip could be used as a hook, a pin, or for cleaning small spaces. Less common responses included using it as a paper airplane weight, as a wire to cut through flowers, or as a token in a game. The individuals who generated the more unusual responses also had a higher number of autism traits. Similarly, when participants were presented with abstract drawings and asked to provide as many ideas as possible to explain the images, those with the highest number of autism traits tended to come up with fewer but more unusual interpretations.

Some people on the autism spectrum have remarkable strengths, and a few have prodigious skill in music, numerical calculation, drawing, and the like. Many of these autistic savants have become well-known. One is Stephen Wiltshire, whom Sir Hugh Casson, former president of the Royal Academy of Arts, considered perhaps the best child artist in Britain. After looking at a building for a few minutes, Wiltshire could draw it quickly, confidently, and accurately. He drew entirely from memory, without notes, and he rarely missed or added a detail. As Casson wrote, "Stephen Wiltshire draws exactly what he sees—no more, no less."[28]

The noted neurologist and author Oliver Sacks was intrigued that Wiltshire could be so gifted artistically despite his enormous emotional and intellectual deficits. This caused him to ask: "Was art not, quintessentially, an expression of a personal vision, a self? Could one be an artist without having a 'self'?"[29] Presumably, anyone who has a sense of self must also have a sense of empathy for others. Sacks worked with Wiltshire over a period of years, and during that time it became more and more evident that the young man had extraordinary perceptual skills, yet never developed great empathy. It was as if the two components of art—the perceptual and the empathic—were separated in his brain.

Another extraordinary artistic savant was Nadia, who at age two and a half started to draw horses and then a variety of other subjects in ways that psychologists thought simply impossible. When she was five years old, she could draw pictures of horses that were comparable to those produced by professionals. She showed an early mastery of space and an ability to depict appearances and shadows, a sense of perspective that gifted child artists don't develop until they are in their teens.[30]

We do not know what accounts for this creativity in autistic people, but a review by Francesca Happé and Uta Frith of numerous studies suggests that superior sensory acuity, focus on detail, visual memory, and detection of patterns may be involved, along with an obsessive need to practice. Almost 30 percent of people on the autism spectrum exhibit special skills in music, memory, numerical and calendar calculations, drawing, or language. Moreover, some individuals have developed multiple talents. Stephen Wiltshire, for example, has perfect pitch and musical talent, as well as his drawing ability. These findings suggest that the biological basis for talent in, say, numerical or calendar calculations does not differ appreciably from the basis of talent in art or music—a conclusion that may extend to neurotypical people as well.[31]

Darold Treffert of the University of Wisconsin, who studies savants, holds that "Serious study of savant syndrome, including the autistic savant, can propel us along further than we have ever been in understanding, and maximizing, both brain function and human potential."[32] Allan Snyder, director of the Centre for the Mind at the University of Sydney in Australia, has pursued the idea that the left hemisphere's control over the creative potential of the right hemisphere is lessened in people with autism.[33]

CREATIVITY IN PEOPLE WITH ALZHEIMER'S DISEASE

Many people with Alzheimer's take up art to communicate with their family. Art thus becomes not only a means of creative expression but also a language they can use when other avenues of communication have failed them.

The converse is also true: artists who develop Alzheimer's disease can continue to paint interesting works. This phenomenon was clearly evident in Willem de Kooning, one of the founders of Abstract Expressionism and the New York School. In 1989 de Kooning was examined and found to have Alzheimer's-like symptoms. He suffered severe memory loss and was often disoriented, but when he entered his studio, he returned to being cogent and engaged. The simplicity, airiness, and lyricism of his later paintings was a dramatic departure from his earlier paintings, and it enriched his body of work.[34] A number of art historians have argued that this should not be surprising, because in many cases, especially among Abstract Expressionists like de Kooning, creativity derives more from intuition than from intellect.

CREATIVITY IN PEOPLE WITH FRONTOTEMPORAL DEMENTIA

When frontotemporal dementia starts on the left side of the brain, it typically affects speech, leading to aphasia. In 1996 Bruce Miller of the University of California, San Francisco, noticed that some of his dementia patients with progressive language disorder had begun to exhibit artistic creativity. People who had painted before started to use bolder colors, and some people who had never painted took up painting for the first time. Specifically, some of Miller's patients with damage to the left frontal regions of their brains were experiencing increased activity in the right posterior regions—regions thought to be involved in creating art.[35]

This outburst of artistic creativity supports John Hughlings Jackson's contention that the left brain and right brain have different functions and that they inhibit each other. Although this distinction oversimplifies the nature of complicated processes such as creativity, which most certainly have multiple origins, we now have enough evidence from imaging studies to conclude that some aspects of artistic and musical creativity do come from the right hemisphere of the brain.

Like Alzheimer's disease, frontotemporal dementia may result in dramatic changes in an artist's style of painting, as well as his or her behavior. In "The Mysterious Metamorphosis of Chuck Close," the writer Wil S. Hylton observes that at age seventy-six the noted painter had radically upended his distinctive style of portraiture—in fact, his entire life. Hylton writes:

> Over the past year, I have been stopping off to see Close in various homes and apartments up and down the Eastern Seaboard, trying to get a handle on the changes in his life and their connection to his work. On my most recent visit to his beach house . . . he looked tan and rested . . . and had been working all morning in the studio behind us on a large self-portrait that I knew he was excited about. . . . [I]t was a radical departure from the last 20 years of his art. Gone were all the swoops and swirls that he typically paints into each square of the grid. In their place, he had filled each cell with just one or two predominant colors, creating a clunky digital effect like the graphics of a Commodore 64. The colors themselves were harsh and glaring, blinding pink and gleaming blue, while the face in the portrait—his face—was cleaved right down the middle, with one side of the canvas painted in different shades from the other.[36]

When Close entered the room and started chatting with Hylton about the painting, he often lost his train of thought. After about a dozen times, Hylton suggested that they take a break, and they agreed to meet again the next day. In thinking about his encounter with Close and his new painting style, Hylton reflected on what the nineteenth-century critic William Hazlitt wrote about the old age of artists: "One feels that they are not quite mortal, that they have one imperishable part about them," what Theodor Adorno called the "late style."[37]

While talking to Hylton the next day, Close mentioned to him that he had received a mistaken diagnosis of Alzheimer's disease the previous year. After spending weeks in a panic, he learned that the diagnosis was wrong and that he had instead another diagnosis.[38] He has since mentioned to others that he has frontotemporal dementia, which would explain both his changed behavior and his brilliant new style.

CREATIVITY AS AN INHERENT PART OF HUMAN NATURE

The idea that creativity is correlated with mental illness is a Romantic fallacy. Creativity does not stem from mental illness; it is an inherent part of human nature. As Rudolf Arnheim points out, "Present psychiatric opinion holds that psychosis does not generate artistic genius but at best liberates powers of the imagination that under normal conditions might remain locked up by the inhibitions of social and educational convention."[39]

Andreasen takes a somewhat different approach to the question of creativity and mental illness. In her essay "Secrets of the Creative Brain," she asks, "Why are so many of the world's most creative minds among the most afflicted?"[40]

To begin with, Andreasen's studies and those of many others support the notion that creativity is not related to IQ. Many people with high IQs are not creative and vice versa. Most creative people are smart, but as Andreasen puts it, they don't have to be "that smart."

What Andreasen did find is that many of the creative writers she studied had suffered from a mood disturbance at some point in their lives, compared to only 30 percent of the controls in her study, who were not as creative as the writers but who had comparable IQ scores. Similarly, Jamison and the psychiatrist Joseph Schildkraut have found that 40 to 50 percent of the creative writers and artists they studied suffered from a mood disorder, whether depression or bipolar disorder.[41]

Andreasen also found that exceptionally creative people were more likely than controls to have one or more first-degree relatives with schizophrenia. This finding suggested to her that some particularly creative people owe their gifts to a subclinical variant of schizophrenia that "loosens their associative links sufficiently to enhance their creativity, but not enough to make them mentally ill."[42]

Andreasen ends her essay on creativity with a quotation from *A Beautiful Mind*, Sylvia Nasar's biography of John Nash, a mathematician who won the Nobel Prize in Economics and who had schizophrenia:

> Nasar describes a visit Nash received from a fellow mathematician while institutionalized at McLean Hospital. "How could you, a mathematician, a man devoted to reason and logical truth," the colleague asked, "believe that extraterrestrials are sending you messages? How could you believe that you are being recruited by aliens from outer space to save the world?" To which Nash replied: "Because the ideas I had about supernatural beings came to me the same way that my mathematical ideas did. So I took them seriously."[43]

In a large study published recently in *Nature Neuroscience*, Robert Power, a scientist affiliated with deCODE Genetics in Iceland, and his colleagues found that genetic factors which raise the risk of bipolar disorder and schizophrenia are more prevalent in people who are in creative professions.[44] Painters, musicians, writers, and dancers were, on average, 25 percent more likely to carry these gene variants than people who work in professions judged to be less creative: farmers, manual laborers, and salespeople. Kári Stefánsson, founder and CEO of deCODE and a coauthor of the study, said: "To be creative, you have to think differently. And when we are different, we have a tendency to be labeled strange, crazy and even insane."[45]

By viewing psychotic states as totally foreign to normal behavior, we fail to recognize that such states are often dramatic representations of character types or temperaments found in the general population—and often found to a greater degree in the minds of creative thinkers, scientists, and artists. That said, people with a brain disorder may very well have readier access to certain aspects of their unconscious than people who are not mentally ill. That difference is particularly critical in terms of creativity. Equally important, the ready accessibility of a mentally ill person to the creativity of his or her unconscious world can be emulated, as Surrealist artists have attempted to show.

LOOKING AHEAD

After setting aside the notion that creativity is inspired by the muses or by madness, and embracing the fact that it is based in the brain, we are nonetheless left with questions.

Creativity feels out of the ordinary to us. We all have an imagination, and we all make creative use of it to solve problems and come up with new ideas. Yet there is something undeniably different about people who are capable of creating remarkable new things. Inner drive and hard work, while essential, don't seem sufficient to explain why some people are extraordinarily creative.

Psychiatric disorders such as schizophrenia and bipolar disorder have illustrated the central role of unconscious mental processes in creativity. Studies of people with autism cast new light on the nature of talent and creative problem solving. Alzheimer's disease and frontotemporal dementia reveal the plasticity of our brain. These disorders may damage the left side of the brain, freeing up the more creative, right side of the brain and resulting in newfound, or radically different, creativity.

What we have learned from biology so far is that creativity results in part from a loosening of inhibitions and the unconscious creation of new associations in the brain. The result is new ways of seeing the world that, Andreasen has found, often occasion strong feelings of joy and excitement.[46] We call on our unconscious in any kind of creative endeavor, whether solving a problem, seeing a slightly new relationship between two scientific findings, painting a portrait, or viewing a portrait.

The unconscious! We call on it in every action, perception, thought, memory, emotion, and decision we make, in sickness and in health. Consciousness is no different. Consciousness is the last great mystery of the human brain, and it, too, as we shall see in chapter 11, entails unconscious processes.

7

MOVEMENT: PARKINSON'S AND HUNTINGTON'S DISEASES

Because movement feels so intuitive to most of us, we may not realize how complicated it is. Before we can act, our brain must issue commands to our body, ordering muscles to flex or relax. Those commands are controlled by the motor system, an elaborate set of neural circuits and pathways that begin in the cortex, extend down the spinal cord, and radiate out to every inch of our body.

When something goes wrong with the motor system, it shows up in unusual behavior or movements, or in loss of control over movement. It also shows up clearly in the brain, which is why neurologists have focused so keenly on anatomy, on tracing neurological disorders to the specific neural circuits in the brain that are responsible for them.

Those studies of neurological disorders contributed greatly to our understanding of normal brain function. In fact, until the 1950s, clinical neurology was known humorously as the medical discipline that could diagnose everything but treat practically nothing. Since then, however, new insights into the molecular underpinnings of neurological disorders have revolutionized treatments for people with Parkinson's disease, stroke, and even severed spinal cords.

Many of the new insights in neurology come from studies of protein folding. Proteins normally fold into specific, three-dimensional shapes. If they misfold or otherwise malfunction, they can clump together in the brain and lead to the death of nerve cells. As we have seen, Alzheimer's disease and frontotemporal dementia are disorders of protein folding. We have now learned that Huntington's, Parkinson's, and other diseases also seem to involve defective protein folding.

In this chapter we begin by examining the workings of the motor system. We then look at what we know about Parkinson's and Huntington's diseases. Finally, we explore the common features of protein-folding disorders, the self-propagation of bizarre proteins known as prions, and genetic studies of protein misfolding.

THE EXTRAORDINARY SKILLS OF THE MOTOR SYSTEM

The motor system controls more than 650 muscles, giving rise to an immense repertoire of possible actions, from the reflexive scratching of an itch to the pirouettes of a ballet dancer, from sneezing to walking a tightrope. Some of these actions are inborn, meaning that our ability to carry them out is built into our brain and spinal cord. Thus, for example, we are programmed to walk upright. But many actions are learned, requiring thousands of hours of practice.

Coordinating all of those muscles is a tremendous challenge, yet the motor system carries out most movements without any conscious instruction. We don't think about how to run or jump or reach for an object, we just do it. How does the brain initiate and coordinate a complex series of actions?

About one hundred years ago the English physiologist Charles Sherrington realized that while our senses provide many ways for information to enter the brain, there is only one way out—movement. The brain takes in a constant barrage of sensory information and ultimately converts it into coordinated movement. If we could understand movement, he reasoned, we would be a giant step closer to understanding the brain.

Sherrington discovered that each of the motor neurons in our spinal cord sends signals to one or more of the body's 650 muscles. Moreover, he realized that in addition to initiating movements and carrying them out, the brain needs feedback about the body's performance. Did the muscle make the intended movement? How quickly? How accurately?

The brain has a special class of neurons that report back on the movement of each muscle. They are known as *sensory feedback neurons*, but they are not the same as the sensory neurons that relay information about the outside world from our sense organs to our brain. The feedback neurons are part of the motor system, and the brain uses information from them to create our internal sense of our own body and the relative position of our limbs in space, a sense known as *proprioception*. Without proprioception, we would be unable to point to an area of our body with our eyes closed or take a step without looking at our feet.

To study the coordinated action of the motor system, Sherrington turned to the simplest motor circuit of all, the reflex. Reflex movements are controlled by a pathway that connects feedback neurons in the muscle directly to motor neurons in the spinal cord—without involving the brain. That's why you can't exert much control over a reflex, even if you try.

By experimenting with reflexes in cats, Sherrington discovered that motor neurons receive and respond selectively to one of two very different signals: *excitatory signals* and *inhibitory signals*. Excitatory signals trigger the action of the motor neurons that initiate extension of a limb, for example, while inhibitory signals tell the motor neurons that control flexion, the opposing movement, to relax. Thus, even a simple knee-jerk reflex requires two simultaneous and opposite commands: the muscles that extend the knee must be excited, while the opposing muscles that flex the knee must be inhibited.

This surprising discovery led Sherrington to formulate a principle that can be applied not just to reflexes but also to the organizational logic of the brain as a whole. In the broadest sense, the task of every circuit in the nervous system is to add up the total excitatory and inhibitory information it receives and determine whether to pass that information along. Sherrington called this principle "the integrative action of the nervous system."[1]

Sherrington demonstrated, for the first time, that we can understand complex neural circuits by studying simpler ones, a principle now widely used in neuroscience. In this sense, he both laid out the challenges that we face today and established a way to overcome them. In 1932 he and Edgar Adrian, whom we met in chapter 1, shared the Nobel Prize in Physiology or Medicine for their discoveries about how neurons orchestrate activity.

PARKINSON'S DISEASE

About 1 million people in the United States have Parkinson's disease. Every year, sixty thousand new cases are detected and a significant number of additional cases evade detection. Worldwide, 7 to 10 million people suffer from this disorder, which usually begins around the age of sixty.

Parkinson's disease was first described in 1817 by the British physician James Parkinson in "An Essay on the Shaking Palsy."[2] Parkinson described six patients, each of whom had three characteristics: tremor at rest, abnormal posture, and slowness and paucity of movement (*bradykinesia*). In time, the patients' symptoms became worse.

It was another century before anything more was published about the disease. In 1912 Frederick Lewy described inclusions, or clumps of proteins, inside certain neurons in the brains of people who had died of Parkinson's disease. Then in 1919 Konstantin Tretiakoff, a Russian medical student in Paris, described the substantia nigra, a part of the brain that he thought was involved in Parkinson's disease (fig. 7.1).

The substantia nigra, or black substance, appears as a dark band on each side of the midbrain. It gets its color from a compound called *neuromelanin*, which we now know is derived from dopamine. What Tretiakoff found during an autopsy on the brain of a person with Parkinson's was decreased pigment, indicating cell loss. Not only that, he saw the inclusions that Lewy had described. Tretiakoff called them Lewy bodies, and they are a hallmark of the disease.

Another forty years went by before Arvid Carlsson discovered dopamine—specifically, low concentrations of dopamine—in the brains of people with Parkinson's disease. Carlsson was interested in three neurotransmitters: noradrenaline, serotonin, and dopamine. He particularly wanted to know which of these was involved in drug-induced Parkinson's. Reserpine, a drug that was used to treat high blood pressure, had been found to cause symptoms of Parkinson's in people and in animals. No one knew how reserpine worked, but early investigators found that it causes a decrease in serotonin.

Carlsson wondered if reserpine also decreased dopamine. He injected the drug into rabbits and found that it makes them listless; their ears droop and they can't move. In an attempt to counteract these effects, he injected the chemical precursor of serotonin into the rabbits. Nothing happened. He then injected the precursor of dopamine, L-dopa, and behold, the animals woke up. Carlsson recognized the importance of his finding, and in 1958 he proposed that dopamine is somehow involved in Parkinson's disease.[3]

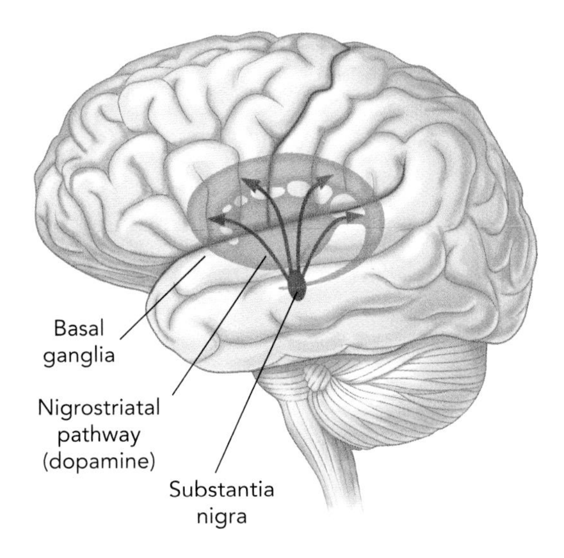

Figure 7.1. Regions of the brain affected by Parkinson's disease. Dopamine produced in the substantia nigra is transmitted along the nigrostriatal pathway to the basal ganglia.

Subsequent studies by Carlsson showed that dopamine is essential to the regulation of muscle movement.[4] As we learned in chapter 4, antipsychotic drugs used to treat people with schizophrenia can reduce dopamine in the brain, resulting in the abnormal muscle movements typical of Parkinson's disease. Carlsson went on to find that the early symptoms of Parkinson's disease result from the death of dopamine-producing neurons in the substantia nigra, although he didn't know then what caused the cell death.[5] Today, we know that those neurons die from a protein-folding disorder: the Lewy bodies inside dopamine-producing neurons are clumps of misfolded proteins that are thought to kill the cells. As the disease worsens, other areas of the brain besides the substantia nigra become involved.

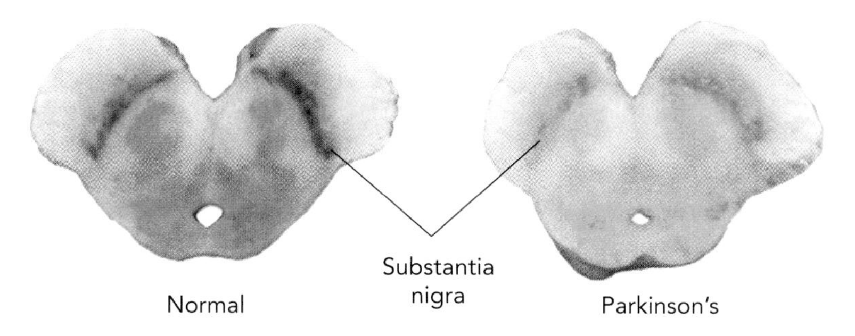

Figure 7.2. People with Parkinson's disease lose dopamine-producing cells (seen as dark patches) in the substantia nigra.

Oleh Hornykiewicz of Austria found at autopsy that dopamine is depleted in the brain of people with Parkinson's (fig. 7.2).[6] In 1967 George Cotzias, of the Brookhaven National Laboratory in New York, gave patients L-dopa to replace the depleted dopamine.[7] Initially, L-dopa was viewed as a cure, but after a honeymoon of several years, it fell out of favor because it was only effective as long as there were dopamine-producing cells in the substantia nigra. It turned out that as more dopamine-producing cells died, the drug's beneficial effects wore off abruptly, leaving patients with involuntary movements, called *dyskinesias.* Clearly, an alternative treatment was needed.

One alternative was surgery. The first effective surgical treatments for Parkinson's disease were undertaken 150 years after Parkinson first described the disorder, by neurosurgeons desperate to help patients with uncontrollable, excessive tremor and limited movement. The surgeons identified, largely by trial and error, specific regions of neural circuits in the basal ganglia and the thalamus that are responsible for the tremor and alleviated their patients' symptoms by destroying those regions.

During the 1970s and '80s, great progress was made in understanding the anatomy and physiology of the motor system, mostly by Mahlon DeLong, then at The Johns Hopkins University and now at Emory University. He found that a particular area of the basal ganglia, the *subthalamic nucleus,* is also rich in dopamine-producing nerve cells and plays an essential role in the control of movement.[8]

Just as DeLong was working on the subthalamic nucleus, a new drug, billed by dealers as "synthetic heroin," showed up on the street. This drug was contaminated with MPTP (1-methyl-4-phenyl-1,2,3,6-tetrahydropyridine), a substance that causes the slowness of movement, tremor, and muscular rigidity typical of Parkinson's disease. After some young people who had taken the drug died, autopsies revealed that MPTP had destroyed the subthalamic nucleus, and with it the brain cells that produce dopamine. Such damage could not be reversed in survivors, but they did respond positively to L-dopa.

Scientists then used MPTP to create a monkey model of Parkinson's disease. They expected to find that the destruction of dopamine-producing cells resulted in underactivity of the subthalamic nucleus, leading to the symptoms of Parkinson's disease. But when DeLong started to record electrical signals from single neurons in the subthalamic nucleus of the monkeys, he found something quite different: the neurons were abnormally active. To his astonishment, the symptoms of Parkinson's disease were caused not by decreased activity of these neurons but by an abnormal increase in activity.

To test whether this abnormal activity was responsible for the tremor and rigidity of Parkinson's disease, DeLong destroyed the subthalamic nucleus in one side of the brain, thus halting the abnormal activity. In 1990 he published the amazing result: damaging the subthalamic nucleus in one side of the brain of a monkey with Parkinson's disease caused the tremor and muscular rigidity on the other side of the body to vanish.[9]

DeLong's discovery led Alim-Louis Benabid, a neurosurgeon at the Joseph Fourier University in Grenoble, France, to start thinking about using deep-brain stimulation to treat people with Parkinson's. Deep-brain stimulation, as we have seen, involves implanting electrodes in the brain and a battery-operated device elsewhere in the body. The device sends high-frequency electrical impulses into a neural circuit, in this case the subthalamic nucleus. The impulses essentially inactivate the circuit, much as the damage to the monkey's subthalamic nucleus did, thus preventing the abnormal activity from interfering with controlled movement (fig. 7.3). The treatment is adjustable and reversible.

By the 1990s, deep-brain stimulation had virtually replaced all other surgical treatments for Parkinson's disease. It does not work for everybody, and it is not a cure: it treats only the symptoms of the disease. If the battery sending electrical impulses should fail or the wires become disconnected, which happens only rarely, the benefit of the treatment is lost almost immediately.

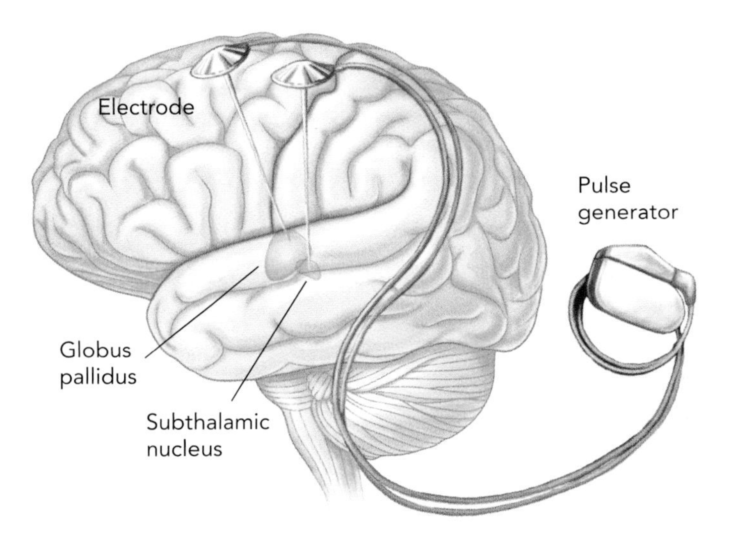

Figure 7.3. Deep-brain stimulation

Deep-brain stimulation has also been used successfully to treat people with psychiatric disorders such as depression. Instead of stimulating the motor circuit to alleviate the symptoms of movement disorders, electrical pulses stimulate the brain's reward system to alleviate the symptoms of depression. Thus deep-brain stimulation may ultimately prove to be a treatment for specific neural circuits rather than for specific diseases.

HUNTINGTON'S DISEASE

Approximately thirty thousand people in the United States have Huntington's disease, a disorder that affects both sexes equally. The age at which the disease first appears varies widely, but the average age of onset is forty. The disorder was first described in 1872 by George Huntington, a Columbia University–trained physician who noted the hereditary nature, involuntary movements, and changes in personality and cognitive functioning that characterize the disorder. His description was so clear and so accurate that other physicians could readily diagnose the disorder, and they named it after him.

Unlike Parkinson's disease, which is fairly localized at first, Huntington's disease can become more widespread quite early and can lead to cognitive as well as motor defects, including sleep disorders and dementia. It primarily affects the basal ganglia, but it also affects the cerebral cortex, the hippocampus, the hypothalamus, the thalamus, and occasionally the cerebellum (fig. 7.4).

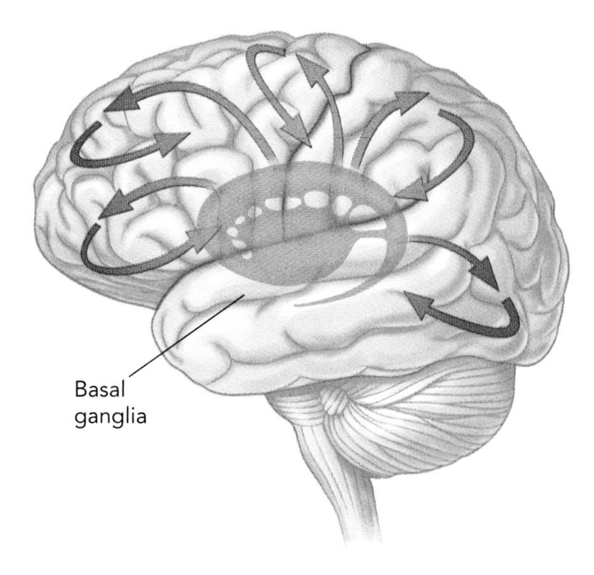

Figure 7.4. Huntington's disease affects the basal ganglia soon after onset and later spreads throughout the cortex.

It took many years to make progress against Huntington's disease, but in 1968 a well-known psychoanalyst—Milton Wexler, whose wife had developed the disease—founded the Hereditary Disease Foundation. Wexler had a dual purpose in mind: to raise funds for basic research and to organize a scientific workforce to focus research on Huntington's disease. This foundation has had a major impact in advancing our understanding of the disease.

Since Huntington's disease is hereditary, the early focus of the foundation was on finding the critical gene. In 1983 David Housman and

James Gusella used a new strategy, called exon amplification, to localize Huntington's disease to a gene on the tip of chromosome 4; they named the gene *huntingtin*.[10]

Ten years later, an international collaborative group called the Gene Hunters, organized by the Hereditary Disease Foundation, finally isolated and sequenced the mutant *huntingtin* gene.[11] Once the gene was isolated, it could be inserted into a worm, a fly, or a mouse to see how the disease would progress. The Gene Hunters noticed that one portion of the *huntingtin* gene is larger than normal. This portion is called a CAG expansion, and it is what causes the disease.

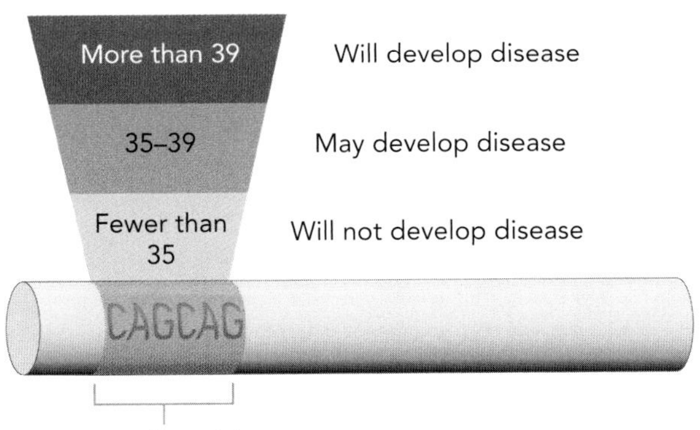

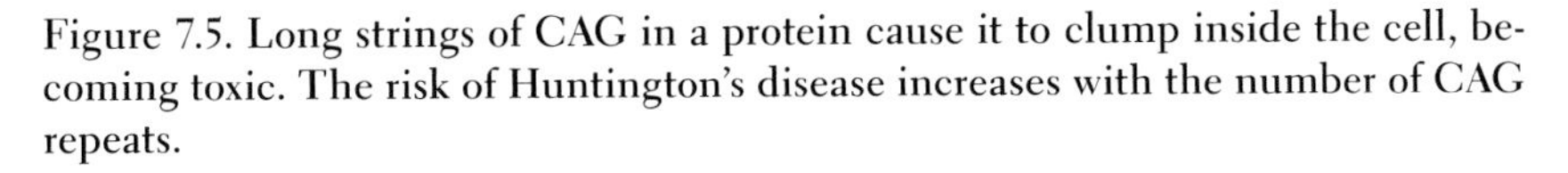

Figure 7.5. Long strings of CAG in a protein cause it to clump inside the cell, becoming toxic. The risk of Huntington's disease increases with the number of CAG repeats.

Our genes are essentially an instruction manual written in a four-letter alphabet: C (cytosine), A (adenine), T (thymine), and G (guanine). Each word is made up of three letters. The word CAG codes for the amino acid glutamine and calls for it to be inserted into a protein when that protein is being synthesized. In Huntington's disease, a portion of the mutant gene repeats the word CAG again and again, resulting in the insertion of too many glutamines. This expanded string of glutamines causes the protein to clump inside the neuron, killing the cell. We all have multiple CAG repeats in this portion of the *huntingtin* gene, but a person who inherits a mutated version of this gene and, as a result, has more than 39 CAGs will develop Huntington's disease (fig. 7.5).

Before long, ten other diseases were discovered to have this CAG expansion, including fragile X syndrome, several distinct forms of spinocerebellar ataxia, and myotonic dystrophy. All of these diseases affect the nervous system, all of them involve misfolded proteins that form clumps, and all of them cause cell death.

COMMON FEATURES OF PROTEIN-FOLDING DISORDERS

We now know that the core molecular cause of Parkinson's disease and Huntington's disease resembles that of several other neurodegenerative disorders: Creutzfeldt-Jakob disease, Alzheimer's disease, frontotemporal dementia, chronic traumatic encephalopathy (the progressive brain degeneration seen in people who have suffered repeated concussions), and the genetic form of amyotrophic lateral sclerosis (ALS, or Lou Gehrig's disease). All of these diseases result from abnormally folded proteins that form clumps in the brain, becoming toxic and eventually killing neurons (fig. 7.6).

In 1982, Stanley Prusiner of the University of California, San Francisco, announced a remarkable discovery: an infectious, abnormally folded protein is involved in Creutzfeldt-Jakob disease, a rare, degenerative brain disorder.[12] Prusiner called this protein a *prion*.

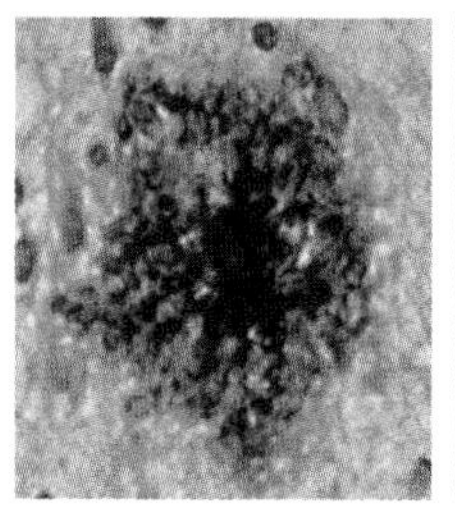

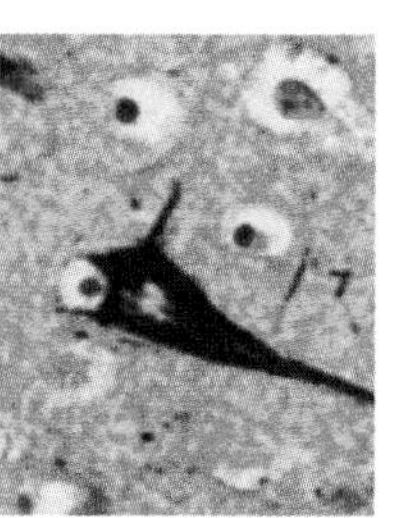

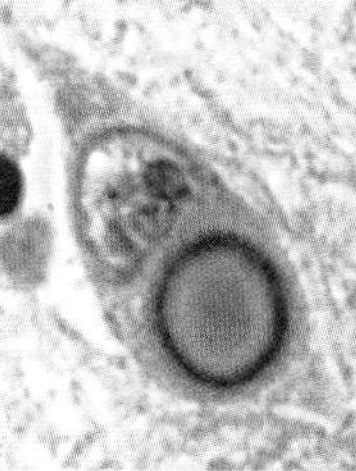

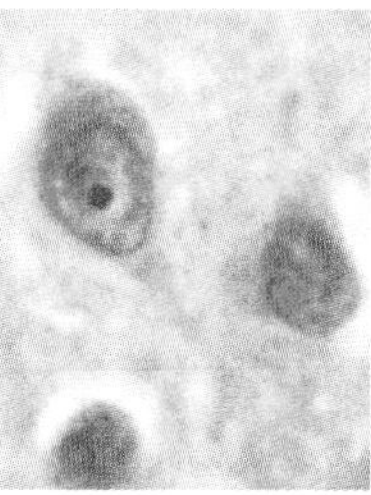

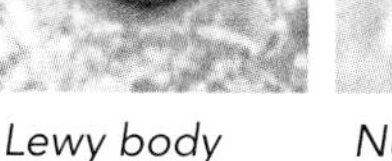

Amyloid-beta plaque	*Tau tangle*	*Lewy body*	*Nuclear inclusion*
Alzheimer's disease	Post-traumatic stress disorder, chronic traumatic encephalopathy	Parkinson's disease	Huntington's disease

Figure 7.6. Abnormally folded proteins form clumps in the brain, leading to neurodegenerative disorders.

Prions are formed when normal precursor proteins misfold. In their normal conformation, precursor proteins mediate healthy cellular functions and are everywhere in the brain. Neurons, like other cells, have internal mechanisms that monitor the shape of proteins. Usually these mechanisms compensate for mutations or damage to the cell, but as we age the mechanisms become weaker and less effective at preventing shape changes. When that happens, a mutant gene or damage to the cell can cause normal precursor proteins to misfold into a lethal prion conformation. The prions form insoluble clumps inside the neuron, disrupting its function and eventually killing it (fig. 7.7).

What makes prions so unusual—and so dangerous—is their ability to self-propagate. In other words, prions do not need genes in order to replicate. As a result, these misfolded proteins are essentially infectious. They can be released by affected neurons and taken up by neighboring cells, where they induce normal precursor proteins to fold abnormally, becoming prions and ultimately killing the cells (fig. 7.8).

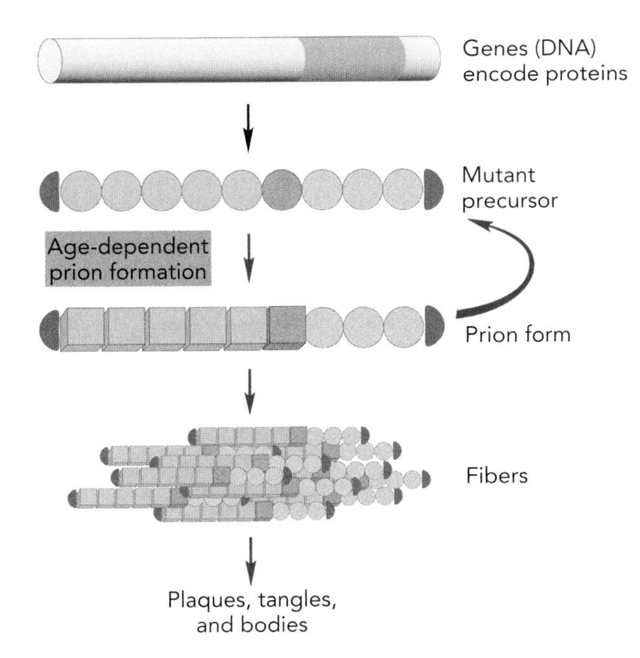

Figure 7.7. Age-dependent prion formation: mutant precursor proteins can cause normal proteins to change shape.

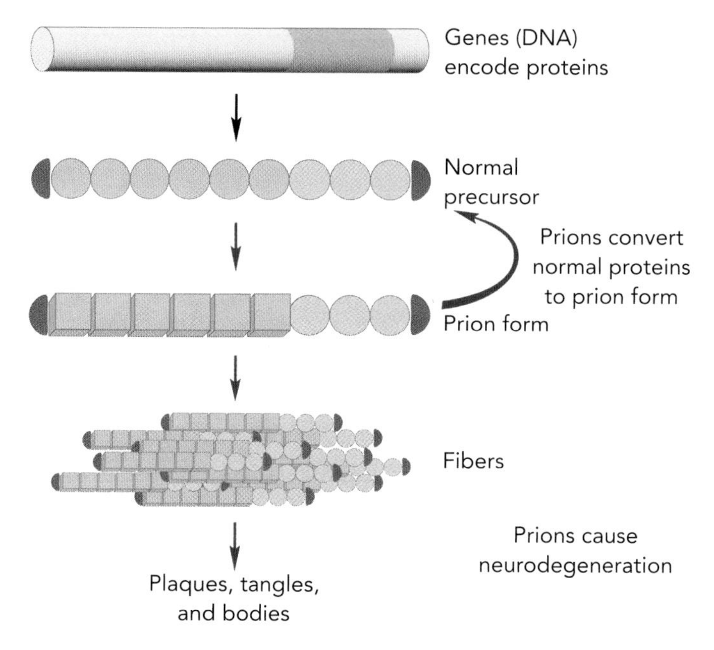

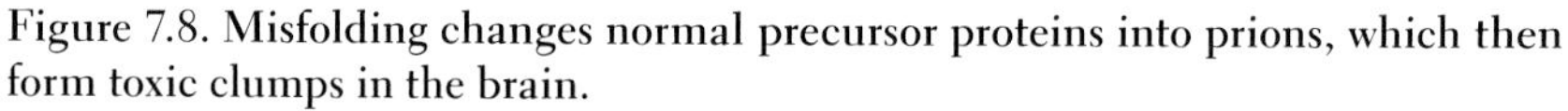

Figure 7.8. Misfolding changes normal precursor proteins into prions, which then form toxic clumps in the brain.

Learning how prions form opened up new possibilities for research directed toward preventing or reversing protein misfolding. Currently, there are no drugs that slow brain degeneration, but prion formation presents three points at which such an intervention might be possible: (1) the point at which a normal precursor protein folds into a prion form, (2) the point at which the prion form aggregates into fibers, and (3) the point at which plaques, tangles, and bodies form (fig. 7.9).

Prusiner's astonishing observations about prions—that they can reproduce and infect other cells, yet contain no DNA—were initially met with considerable resistance in many scientific quarters. But in 1997, fifteen years after discovering these self-replicating, misfolded proteins, Prusiner was awarded the Nobel Prize in Physiology or Medicine. In 2014 he wrote a book about his experiences during those years:

> I wrote this book because I feared that neither science historians nor journalists could construct an accurate narrative of my investigations. This is a first-person account of the thinking, the experiments, and the surrounding events that led to the identification

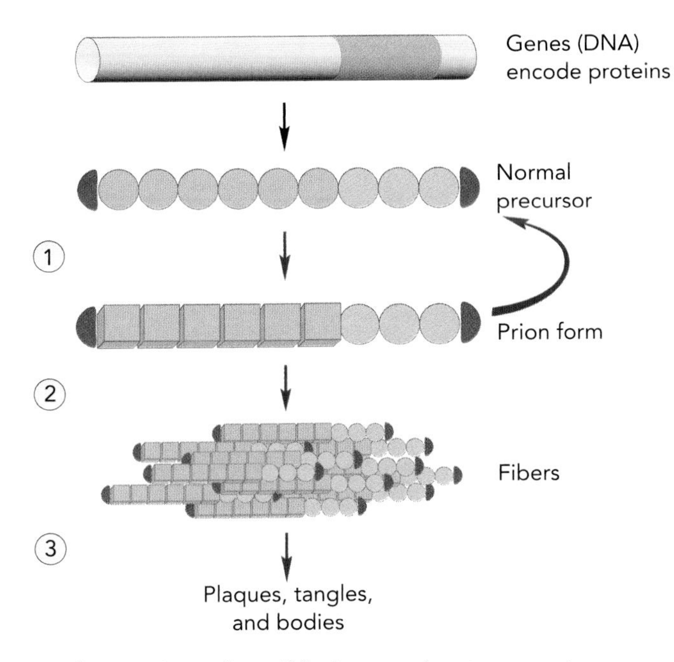

Figure 7.9. Three points of possible intervention to prevent or reverse protein misfolding

of infectious proteins, or "prions" as I named them. I have tried to describe what appears in retrospect to be an audacious plan to define the composition of the agent that causes scrapie, a barnyard disease whose etiology was a mystery at the time. On many occasions, I worried that my data might lead me down dead-end paths. Despite my fascination with the problem, I was haunted by a fear of failure; my anxiety was palpable at almost every turn. Was the problem intractable? As small successes emerged, so did a legion of naysayers who questioned both the wisdom of my pursuit and my scientific prowess; indeed, there were times when little but my naïveté and exuberance sustained me.

The skeptical and frequently hostile reactions to prions from many precincts of the scientific community reflected resistance to a profound change in thinking. Prions were seen as an anomaly: they reproduce and infect but contain no genetic material—neither DNA nor RNA; thus they constitute a disruptive transition in our understanding of the biological world. The consequences of the prion discovery are immense, and they continue to expand. Their causative role in Alzheimer's and Parkinson's diseases has important implications for the diagnosis as well as the treatment of these common, invariably fatal maladies.[13]

GENETIC STUDIES OF PROTEIN-FOLDING DISORDERS

Drosophila, the fruit fly, is the invertebrate animal model par excellence. It was first developed as an experimental organism by Thomas Hunt Morgan at Columbia University to study the basic function of chromosomes in heredity. Later, Seymour Benzer focused on genes that are involved in behavior. He found that genes work together in complex networks called *gene pathways*.

In many diseases, fruit flies and people share not just genes but entire gene pathways. Scientists use these shared features, which have been conserved through the course of evolution, to gain crucial insights into human disease, including neurological disorders. One advantage of using flies is that it speeds up the research process. A disease like Parkinson's can take decades to appear in people, but it will take only days or weeks to appear in flies. A key gene that is mutated in Parkinson's disease, *synuclein alpha*, or *SNCA*, was first identified in the fruit fly (fig. 7.10).

Parkinson's disease usually occurs spontaneously, for reasons that are still not known, but several factors play a role, including the patient's genes (certain gene variants are thought to increase the risk of Parkinson's disease) and exposure to certain toxins. In its rare inherited forms, the *SNCA* gene is mutated, resulting in excessive amounts of the alpha-synuclein protein in the brain, in misfolded alpha-synuclein proteins in the brain, or both. Since all Parkinson's patients, even those who did not inherit the disease, have one or both of these protein abnormalities in the brain, scientists concluded that the mutated gene might reveal some general aspect of the disease.

It turns out that the protein produced by the mutated gene is the main component of Lewy bodies. These bodies are the toxic clumps that form inside neurons when the alpha-synuclein protein folds abnormally.

Researchers inserted the mutated *SNCA* gene into the dopamine-producing neurons of the fruit fly brain to see what would happen. They knew that dopamine is essential to muscle control and that insufficient dopamine causes the palsy and other abnormal movements characteristic of Parkinson's disease. The scientists found that by inserting the mutated gene, they had compromised the dopamine-producing neurons' ability to function. The result was behavioral effects in flies that are strikingly similar to the effects of Parkinson's disease in people.[14]

Flies, like people, have conserved molecular pathways—called *molecular chaperone pathways*—that help proteins take on their normal shape and that sometimes even reverse misfolding. By helping proteins to fold properly, the chaperone pathways prevent clumping. Scientists wondered what would happen if they gave the flies more of the helper proteins that act in these pathways. Perhaps the presence of more helper

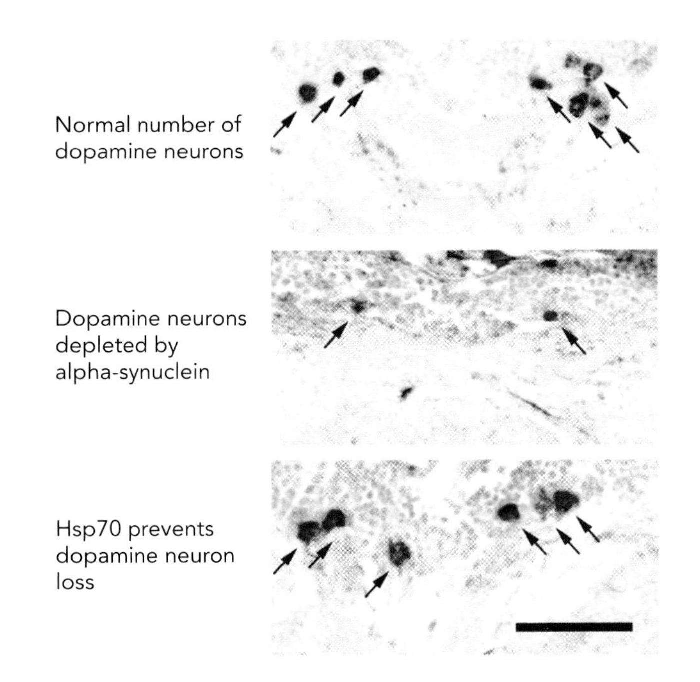

Figure 7.10. The brain of the fruit fly with normal alpha-synuclein protein (*top*); alpha-synuclein protein produced by a mutated gene (*center*); and the mutant protein with the helper protein Hsp70, which promotes normal refolding (*bottom*). Dopamine-producing neurons are indicated by arrows.

proteins would encourage the normal folding of alpha-synuclein proteins and the healthy production of dopaminergic neurons.

By adding helper proteins, the dopamine-producing neurons were no longer compromised. Chaperone proteins have also been found to protect against movement disorders: flies with a mutant *SNCA* gene are poor climbers, but when flies with the same mutation overexpress chaperone proteins they are able to climb normally. This technique also works in fruit fly models of other neurodegenerative diseases—of which there are many now—as well as in mouse models of some neurodegenerative diseases, illustrating once again the utility of animal models for the study of human disease.

LOOKING AHEAD

Parkinson's and Huntington's diseases, Alzheimer's and frontotemporal dementia, Creutzfeldt-Jakob disease, and chronic traumatic encephalopathy all produce widely varying effects on our thinking and behavior, our memory and emotions. Yet we now know that these and other neurodegenerative disorders share an underlying molecular mechanism: the failure of proteins to fold correctly, thus eventually killing neurons.

We also know that the function of any given protein is determined by its unique shape, a shape arrived at through an extraordinarily precise process of folding. Thus, the dramatically different symptoms caused by protein-folding disorders are attributable to changes in the shape of particular proteins responsible for particular functions in the brain. As we have seen, the death of dopamine-producing neurons, caused by misfolded proteins, leads to Parkinson's disease. A mutant gene that orders up too many glutamines during protein synthesis results in misfolded proteins that clump in the brain and cause Huntington's disease, as well as several other diseases of the nervous system. The self-replicating, misfolded proteins known as prions, which are responsible for the toxic tangles found in Creutzfeldt-Jakob and related diseases, can even act as infectious agents.

At present, there are no drugs that slow brain degeneration, although deep-brain stimulation can calm the neural circuits responsible for uncontrolled movement, thereby giving relief to people with Parkinson's disease. Research on neurological disorders now includes genetic and molecular studies, which may provide scientists with points of entry for preventing or reversing the process of protein misfolding. As we have seen, genetic studies in animal models are already beginning to move us toward that goal.

8

THE INTERPLAY OF CONSCIOUS AND UNCONSCIOUS EMOTION: ANXIETY, POST-TRAUMATIC STRESS, AND FAULTY DECISION MAKING

When we shop in a supermarket or chat with strangers at a party, we unconsciously rely on our emotions to help us navigate the situation. We also rely unconsciously on our emotions when we make decisions. Emotions are states of readiness that arise in our brain in response to our surroundings. They give us critical feedback about the world and set the stage for our actions and decisions. In chapter 3, we considered emotion in the context of mood, our individual temperament—specifically, we considered what the biology of mood disorders has revealed about our sense of self. In this chapter, we examine the nature of emotion—its conscious and unconscious components—and the essential role it plays in other aspects of our life.

Our brain has an approach-avoidance system that encourages us to seek out experiences that evoke pleasurable emotions and to avoid those that evoke painful or frightening ones. In this chapter we explore what studies of animals have taught us about how the brain regulates the emotion of fear, and we look at the nature of human anxiety disorders, particularly post-traumatic stress disorder, which is an extreme reaction of fear. By studying these disorders, scientists are learning where emotions arise in the brain and how they control our behavior. We learn about novel ways in which scientists are using drug therapy and psychotherapy to help treat people with anxiety disorders.

Because emotions are a powerful force in any decision we make, from the simplest to the most complex, this chapter touches on important biological aspects of how we reach decisions, including moral decisions. We see how damage to regions of the brain that regulate emotion damps down our emotions and adversely affects our ability to make choices, and we see how deficits in the brain regions that control emotional processing and moral decision making can lead to psychopathic behavior.

THE BIOLOGY OF EMOTION

The first person to consider the biology of emotion was Charles Darwin. In the course of his work on evolution, Darwin came to understand that emotions are mental states shared by all people in all cultures. He was particularly interested in children because he believed that they express emotions in a pure and powerful form. Since they are seldom able to suppress their feelings or fake an expression, he considered them ideal subjects for studying the importance of emotion (fig. 8.1). In his 1872 book, *The Expression of the Emotions in Man and Animals*, Darwin also carried out the first comparative study of emotion across species. He showed that unconscious aspects of emotion are present in animals as well as people and noted that these unconscious aspects have been extremely well conserved throughout evolution.

We are all familiar with emotions such as fear, joy, envy, anger, and excitement. To some extent these emotions are automatic: the brain systems that carry them out operate without our being aware of them. At the same time, we experience feelings of which we are fully aware, so that we are capable of describing ourselves as scared or angry or grumpy, surprised or happy. The study of emotions and moods helps reveal the porous boundaries between unconscious and conscious mental processes, documenting the ways in which these seemingly distinct kinds of cognition are constantly interacting. We first encountered the divide between unconscious and conscious processing in the brain when we explored creativity, in chapter 6, and we will return to it again when we discuss the unconscious, in chapter 11.

All of our emotions have two components. The first begins uncon-

Figure 8.1. Darwin studied emotion in children because they display emotions in their purest form.

sciously and manifests itself as an outward expression; the second is a subjective, internal expression. The great American psychologist William James described these two components in an 1884 essay entitled "What Is an Emotion?" James had a profound insight: not only does the brain communicate with the body, but, equally important, the body communicates with the brain.

James proposed that our conscious experience of emotion takes place only *after* the body's physiological response, that the brain responds to the body. He argued that when we encounter a potentially dangerous situation, such as a bear in our path, we do not consciously evaluate the danger and then feel afraid. Instead, we respond instinctively and unconsciously to the sight of the bear by running away from it and only later experience fear. In other words, we process emotion first from the bottom up—with a sensory stimulus that causes our heart rate and respiration to spike, leading us to flee—and only then from the top down—using cognition to explain the physiological changes that have taken place in our body. James noted that "without the bodily states following on the perception, the latter would be purely cognitive in form, pale, colorless, destitute of emotional warmth."[1]

The second component of emotion is the subjective, internal experience of emotion, the conscious awareness of how we feel. In this book, we follow the lead of Antonio Damasio, director of the Brain and Creativity Institute at the University of Southern California, and restrict the word "emotion" to the observable, unconscious behavioral component and use "feeling" to refer to the subjective experience of emotion.

THE ANATOMY OF EMOTION

Emotions can be classified along two axes: *valence* and *intensity*. Valence has to do with the nature of an emotion, with how bad or good something makes us feel on a spectrum from avoidance to approach (fig. 8.2). Intensity refers to the strength of the emotion, the degree of arousal it evokes (fig. 8.3). We can actually map most emotions onto these two axes. Such a map doesn't capture the entire essence of a particular emotion, but it does present it in a way that is useful when matching facial expressions to the brain systems that produce them.

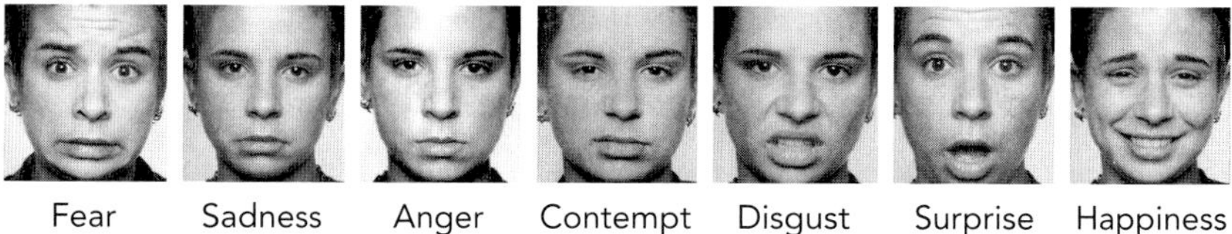

Figure 8.2. The valence of emotion, from avoidance to approach

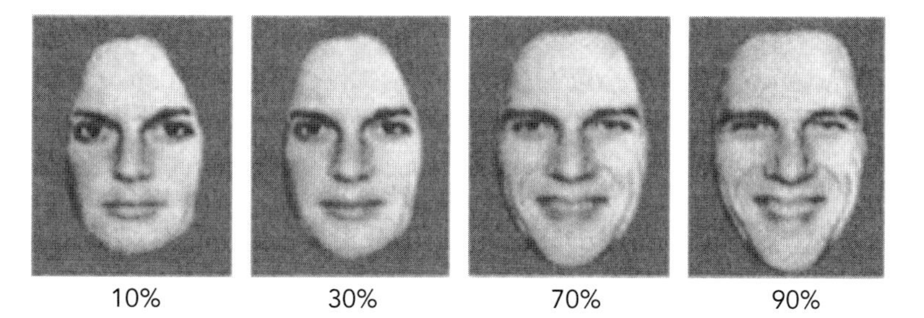

Figure 8.3. The intensity spectrum of happiness

Many structures in the brain are involved in emotion, but four of them are particularly important: the hypothalamus, which is the executor of emotion; the amygdala, which orchestrates emotion; the striatum, which comes into play when we form habits, including addictions; and the prefrontal cortex, which evaluates whether a particular emotional response is appropriate to the situation at hand (fig. 8.4). The prefrontal cortex interacts with, and in part controls, the amygdala and striatum.

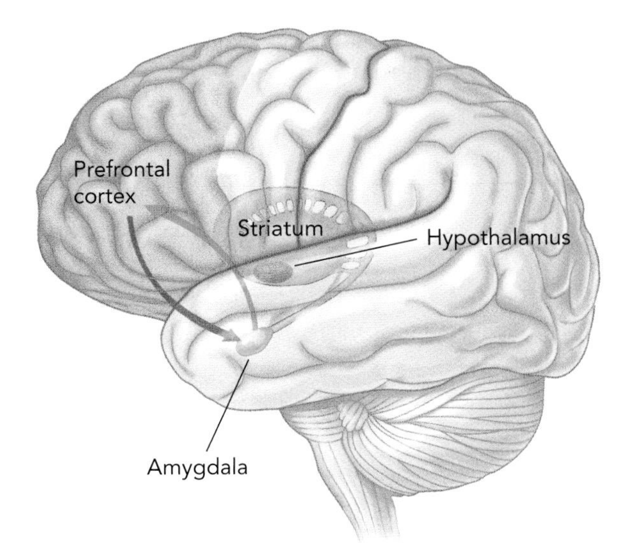

Figure 8.4. The hypothalamus, amygdala, striatum, and prefrontal cortex are the four main structures in the brain involved in emotion.

We say the amygdala "orchestrates" emotion because it links the unconscious and conscious aspects of an emotional experience. When the amygdala receives sensory signals from the areas concerned with vision, hearing, and touch, it generates responses that are relayed onward, largely by the hypothalamus and other structures in the brain that control our automatic physiological responses. When we laugh or cry—when we experience any emotion—it is because these brain structures are responding to the amygdala and acting on its instructions. The amygdala is also connected to the prefrontal cortex, which regulates the feeling state, the conscious aspects of emotion, and its influence on cognition.

It goes without saying that our emotions need to be regulated. Aristotle argued that the proper regulation of the emotions was a defining feature of wisdom. "Anyone can become angry—that is easy," he wrote in *The Nicomachean Ethics*. "But to be angry with the right person and to the right degree and at the right time and for the right purpose, and in the right way—that is not within everybody's power and is not easy."[2]

FEAR

Like every other emotion, fear has both an unconscious and a conscious component. The physical aspects of our emotional response to a fearful stimulus—accelerating heart rate and respiration and opening sweat glands in the skin—are mediated by the autonomic nervous system, and they take place below the level of consciousness. As we have seen, James argued that our bodily response to fear comes first and triggers our conscious feeling. Thus, without the body there would be no fear. This insight set the agenda for the study of fear.

Scientists have a very good understanding of the neural circuitry of fear. It begins with the amygdala, which orchestrates all emotion but seems to be particularly sensitive to fear. A scary stimulus arrives at the amygdala, activates a representation of danger, and triggers the body's fear response. These are automatic, hardwired physiological and behavioral responses.

Next in the circuit is the insular cortex, a small island of neurons lying deep within the frontal and parietal lobes that translates bodily emotion into conscious awareness. It assesses bodily responses, such as degree of pain, and monitors what is going on in the viscera and the muscles, assiduously tracking our heart rate and the activity of our sweat glands. The later discovery of the insular cortex provided biological confirmation of James's idea that our bodily response to fear precedes our awareness of fear.

Another region involved in the neural circuitry of fear—and of anger—is a part of the prefrontal cortex known as the ventromedial prefrontal cortex. This structure is also important for what we would call moral emotions—indignation, compassion, embarrassment, and shame.

Finally, a second region of the prefrontal cortex, the dorsal prefrontal cortex, is actually the point at which our conscious mind—our volition, or will—can impose itself on the way emotion is being carried out.

Our reaction to fear is an *adaptive response*, one that helps us survive. It is a program of actions sometimes referred to as the "fight, flight, or freeze" response. These actions include musculoskeletal changes (the facial muscles assume a mask of fear), changes in posture (a sudden startled movement, followed by rigidity), increases in heart rate and respiration, contraction of the stomach and intestinal muscles, and secretion of stress hormones such as cortisol. All of these changes in the body take place in concert, and they send signals to the brain.

Two things about fear are important here. First, the senses send signals to the amygdala, which recruits additional areas of the brain. We know this because brain imaging gives us a precise portrait of what is happening as this primal response unfolds. Second, the changes in our body, in concert with the insular cortex, make us aware of the feeling. We feel scared because the brain has noticed the changes unfolding within our body. That is why we get ready to run before we know why we are running.

THE CLASSICAL CONDITIONING OF FEAR

Until the end of the nineteenth century, the only approaches to the mysteries of the human mind were introspection, philosophical inquiries, and the insights of writers. Darwin changed all that when he argued that human behavior evolved from our animal ancestors. This argument gave rise to the idea that experimental animals could be used as models to study human behavior.

The first person to explore this idea systematically was Ivan Pavlov, who had won the Nobel Prize in Physiology or Medicine in 1904 for his study of gastric secretion. As we saw in chapter 5, Pavlov taught dogs to associate two stimuli—a neutral stimulus (such as the sound of a bell) that predicts a reward (or punishment) and a positive (or negative) reinforcing stimulus. These experiments showed that the brain is able to recognize and make use of a stimulus to predict an event (the arrival of food) and to generate a behavior (salivation) in response.

Pavlov used this finding not only to study positive reinforcement, the anticipation of something pleasurable, but also to study negative reinforcement, the consequences of fear. He did this by pairing a neutral stimulus (the sound of a bell) with an electric shock. Not surprisingly, applying an electric shock to the feet of a dog causes the animal to manifest intense fear. We cannot say what the dog is feeling—we have no way of asking it—but we can observe the dog's behavior, its expression of fear.

Joseph LeDoux, a neuroscientist at New York University, adapted Pavlov's strategy to rats and mice.[3] He put an animal in a small chamber and sounded a tone. The animal simply ignored the tone. Then, instead of sounding a tone, LeDoux shocked the animal. This time, it responded by jumping and flinching. Finally, LeDoux sounded the tone just before he administered a shock. The animal soon associated the tone with the shock—that is, it learned that the tone predicted the shock. The next time the animal heard the tone, whether the next day, two weeks later, or a year later, it responded with the classic fear response: it froze in its cage, and its blood pressure and heart rate skyrocketed.

The fear response results from the association of the tone and the shock. As we have seen, all sensory information related to emotion travels into the brain via the amygdala. A sound, for example, goes first to the auditory thalamus; from there it is relayed directly to the amygdala and indirectly to the auditory cortex (fig. 8.5). In other words, a sound reaches the amygdala and activates the fear response before it reaches the auditory cortex. The direct pathway to the amygdala is quick, but the

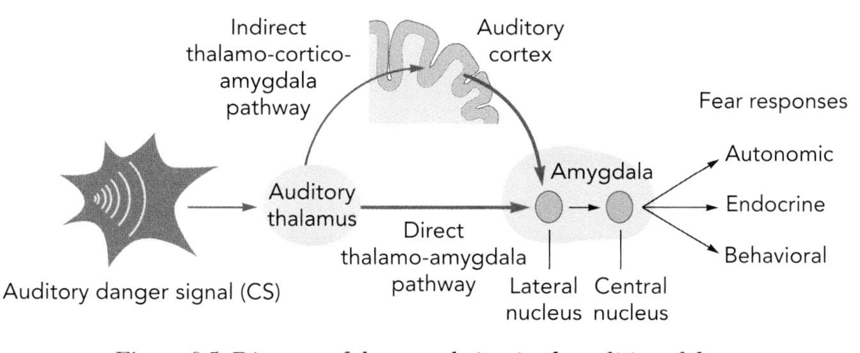

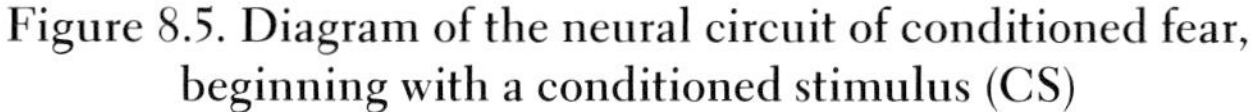
Figure 8.5. Diagram of the neural circuit of conditioned fear, beginning with a conditioned stimulus (CS)

information it carries is not very precise. That is why the sound of a car backfiring frightens us—until we realize what the sound is.

How does this learning take place within the amygdala? One of the key requirements scientists have discovered is that, for a fear association to be created, stored, and consolidated in the brain, the tone and the shock must give rise to classical conditioning. Classical conditioning takes place when the tone and the shock are registered sequentially (tone closely followed by shock) by the same cells in the lateral nucleus, the first relay area of the amygdala. When that happens, the tone, which was initially ineffective in activating those cells, becomes highly effective, causing them to send information to the central nucleus of the amygdala. The central nucleus activates motor cells and thus initiates action—jumping and flinching—in response to the sound.

Because two areas of the amygdala are involved in fear, scientists have come to understand that people can develop pathological fear in two different ways. In some people, the lateral nucleus has learned to be overly sensitive to the world, responding with fear to things that others don't even notice, such as people walking by or the sounds of a bird flying overhead. In other people, the central nucleus is overly reactive, triggering emotional responses that are disproportionate to the threat.

Research on the anatomy of the fear response—on how rodents react to a shock—has deepened our understanding of how people respond to fear. When the circuits of fear in our brain go awry, they give rise to various anxiety disorders. Imaging studies have confirmed that the amygdala is hyperactivated in people who are coping with anxiety, post-traumatic stress, and other fear-related disorders.

HUMAN ANXIETY DISORDERS

We all become anxious occasionally, especially when confronted by danger. But if we experience a chronic state of excessive worry and guilt for no discernible reason, we are suffering from a generalized anxiety disorder. These disorders frequently occur with depression. Fear-related anxiety disorders include panic attacks, phobias (such as fear of heights, animals, or public speaking), and post-traumatic stress disorder. For many years, the various anxiety disorders were considered separate syndromes, but because of their similarities, scientists now regard them as a related cluster of disorders.

Nearly one-third of all Americans will experience symptoms of an anxiety disorder at least once during their lifetime, making these disorders the most common psychiatric illnesses by far. Moreover, anxiety disorders can affect children as well as adults.

Perhaps the most widely known fear-related disorder is post-traumatic stress disorder (PTSD), which is caused by experiencing or observing life-threatening events such as physical assault or abuse, war, terrorist attack, sudden death, or natural disaster. All told, about 8 percent of the U.S. population—at least 25 million people—will experience PTSD at some point in their lifetime. More than forty thousand U.S. war veterans are known to be affected by the disorder, and thousands more cases are thought to be unreported (fig. 8.6).

Exposure to trauma affects the amygdala, which generates our re-

Figure 8.6. Post-traumatic stress disorder has afflicted soldiers throughout history. A marine returns after two days of battle on the beaches of the Marshall Islands in February 1944.

sponse to fear, and the dorsal prefrontal cortex, which helps regulate our response to fear, but trauma is especially damaging to the hippocampus. The hippocampus, as we have seen, is critical for storing memories of people, places, and objects, but it is also important for recalling memories in response to environmental stimuli. As a result of trauma's damage to the hippocampus, people with PTSD experience several major symptoms: they have flashbacks, or spontaneous re-experiencing of the traumatic event; they avoid sensory experiences associated with the initial event; they become emotionally numb and withdraw from others; and they are irritable, jumpy, aggressive, or have trouble sleeping. The disorder is commonly accompanied by depression and substance abuse, and can lead to suicide.

Most psychiatric disorders, as we have seen, involve the interaction of a genetic predisposition with an environmental trigger. Post-traumatic stress disorder is a perfect example of this interaction. Not everyone who is exposed to a traumatic stress will develop PTSD. In fact, if one hundred people were exposed to the same traumatic event, about four men and ten women would develop the disorder. (Scientists don't know why men who experience traumatic stress are so much less likely to develop PTSD.) In addition, studies of identical twins suggest that if one twin responds to a trauma with PTSD, the other twin will also develop PTSD in response to that trauma. These findings indicate that one or more genes predispose people to the disorder, and this may also explain why PTSD so often occurs with other psychiatric disorders: they may share some of the same genes.

Another primary cause of PTSD is childhood trauma. People who have suffered trauma as children are much more likely to develop PTSD as adults because trauma affects the developing brain differently than it does the adult brain. Notably, early trauma can cause *epigenetic changes*, that is, molecular changes in reaction to the environment that do not alter the DNA of a gene but do affect the expression of that gene. Some of these epigenetic changes are initiated in childhood and persist into adulthood. One such change is known to occur in a gene that regulates our response to stress; this change heightens the risk of developing PTSD in response to traumatic stress in adulthood.

TREATING PEOPLE WITH ANXIETY DISORDERS

At present, the two main classes of treatment for anxiety disorders are medication and psychotherapy. Both decrease activity in the amygdala, but they do so in different ways.

As we learned in chapter 3, depression is commonly treated with drugs that increase the concentration of serotonin in the brain. The same antidepressants are effective in treating 50 to 70 percent of people with generalized anxiety disorders because they lessen worry and guilt, feelings that are associated with depression. However, the drugs do not work nearly as well for people with specific fear-related disorders. For them, psychotherapy has proven much more effective. PTSD, for example, can be managed with cognitive behavioral therapy, including *prolonged exposure therapy* and *virtual reality exposure therapy.*

Recently, Edna Foa and others have shown that prolonged exposure therapy works particularly well for people with fear-related disorders.[4] This form of psychotherapy essentially teaches the brain to stop being afraid by reversing learned fear associations in the amygdala. If we were to try to extinguish fear in LeDoux's mice, for example, we would present the animals with the tone over and over again—but without the electric shock. Eventually, the synaptic connections underlying the fear association would weaken and disappear, and the mice would no longer cringe in response to the tone.

While exposing a person to the cause of his or her fear only a few times can actually exacerbate fear, proper use of exposure therapy can extinguish or inhibit it. Sometimes, this involves exposing patients to a virtual experience. Virtual experiences are useful in situations that might be difficult in real life, such as riding an elevator a hundred times. Results produced by virtual exposure are almost as effective as their real-world counterparts.

Barbara Rothbaum, director of the Trauma and Anxiety Recovery Program at Emory University, is a pioneer in virtual reality exposure therapy. She began by fitting Vietnam veterans who had chronic PTSD with a helmet that plays one of two filmed scenarios: a landing zone or the inside of an in-flight helicopter. She then followed the patients' reactions on a monitor and talked to the patients as they re-experienced traumatic events. When this therapy proved effective, she extended it to other patients as well.[5]

Another approach is to erase a terrifying memory entirely. As we learned in chapter 5, short-term memory results when existing connections among synapses are strengthened, but long-term memory requires repeated training and the formation of new synaptic connections. In the interim, while a memory is being consolidated, it is sensitive to disruption. Recent studies have revealed that a similar sensitivity to disruption occurs when a memory is retrieved from long-term storage; that is, memories become unstable for a short period of time after they have been retrieved.[6] Thus, when a person recalls a memory that evokes the fear response (or, in the case of a rat, when it is re-exposed to the tone), the memory is destabilized for several hours. If during that time the storage processes in the brain are perturbed, either behaviorally or with a drug, the memory often does not go back into storage properly. Instead, it is erased or made inaccessible. Thus the rat is no longer afraid, and the person feels better.

Alain Brunet, a clinical psychologist at McGill University in Montreal, studied nineteen people who had been suffering for several years from PTSD.[7] (Their traumas included sexual assaults, car crashes, and violent muggings.) People in the treatment group were given propranolol, a drug that blocks the action of noradrenaline, a neurotransmitter released in response to stress that triggers our fight, flight, or freeze response. Brunet gave one group of study participants a dose of propranolol, then asked them to write a detailed description of their traumatic experience. While the participants were remembering the awful event, the drug suppressed the visceral aspects of their fear response, thereby containing their negative emotions. As James was the first to suggest, minimizing the body's emotional response can also minimize our conscious awareness of the emotion.

One week later, the patients returned to the lab and were asked to remember the traumatic event once again. Participants who had not received propranolol exhibited high degrees of arousal that were consistent with anxiety (for example, their heart rate spiked suddenly), but those given the drug had significantly lower stress responses. Although they could still remember the event in vivid detail, the emotional component of the memory located in the amygdala had been modified. The fear wasn't gone, but it was no longer crippling.

Emotions do more than affect our behavior; they also affect the decisions we make. We accept that we sometimes make hasty decisions in response to our feelings. But surprisingly, emotion plays a role in *all* of our decisions, even moral ones. In fact, without emotion, our ability to make sound decisions is impaired.

EMOTION IN DECISION MAKING

William James was one of the first scientists to propose a role for emotion in decision making. In his 1890 textbook, *The Principles of Psychology*, he launched into a critique of the "rationalist" account of the human mind. "The facts of the case are really tolerably plain," he wrote. "Man has a far *greater* variety of impulses than any other, lower animal."[8] In other words, the prevailing view of humans as purely rational creatures, defined "by the almost total absence of instincts," was mistaken. James's principal insight, however, was that our emotional impulses aren't necessarily bad. In fact, he believed that the preponderance of habits, instincts, and emotions in the human brain is an essential part of what makes our brain so effective.

Scientists have recorded several powerful demonstrations of the importance of emotion in decision making. In his book *Descartes' Error*, Antonio Damasio describes the case of a man named Elliot.[9] In 1982 a small tumor was discovered in the ventromedial prefrontal cortex region of Elliot's brain. The tumor was removed by a team of surgeons, but the resulting damage to his brain changed his behavior dramatically.

Before the operation, Elliot had been a model father and husband. He held an important management job in a large corporation and was active in his local church. After the surgery, Elliot's IQ stayed the same—he still tested in the ninety-seventh percentile—but he exhibited several profound flaws in decision making. He made a series of reckless choices and started a series of businesses that quickly failed. He got involved with a con man and was forced into bankruptcy. His wife divorced him. The IRS began investigating him. Eventually, he had to move in with his parents. Elliot also became quite indecisive, especially when it came to minor details such as where to eat lunch or what radio station to listen to. As Damasio would later write, "Elliot emerged as a man with a normal intellect who was unable to decide properly, especially when the decision involved personal or social matters."[10]

Why was Elliot suddenly incapable of making good personal decisions? Damasio's first insight occurred while talking to Elliot about the tragic turn his life had taken. "He was always controlled," Damasio writes, "always describing scenes as a dispassionate, uninvolved spectator. Nowhere was there a sense of his own suffering, even though he was the protagonist. . . . I never saw a tinge of emotion in my many hours of conversation with him: no sadness, no impatience, no frustration."[11]

Intrigued by this emotional deficit, Damasio hooked Elliot up to a machine that measures the activity of the sweat glands in the palms of the hands. (Whenever we experience strong emotions, our skin is literally aroused and our palms start to perspire.) Damasio then showed him various photographs that would normally trigger an immediate emotional response: a severed foot, a naked woman, or a house on fire. No matter how dramatic the picture, Elliot's palms never got sweaty. He felt nothing. Clearly, the surgery had damaged an area of the brain that is essential for processing emotion.

Damasio began to study other people with similar patterns of brain damage. They all appeared perfectly intelligent and showed no deficits on any conventional cognitive tests, yet they all suffered from the same profound flaw: they didn't experience emotion and therefore had tremendous difficulty making decisions.

MORAL DECISION MAKING

The first indication of a link between moral functions and the brain dates back to 1845 and the famous case of Phineas Gage, which we touched on in chapter 1. Gage, a railroad worker, was handling explosives when a terrible accident occurred: an iron bar was driven through his skull. The bar entered the base of his skull and came out at the top, damaging his brain severely (fig. 8.7). A local physician took excellent care of him, and Gage recovered physically to an amazing degree. Within days, he was able to walk and talk and function effectively. Within a few weeks he was back at the job. But Gage had changed dramatically.

Before the accident, Gage had been the foreman of the crew. He was absolutely reliable. He could always be counted on to do the job and do it well. After the accident, he was completely irresponsible. He never showed up on time. He became obscene in his language and his behavior. He paid no attention to his fellow workers. He had lost any sense of moral judgment.

Many years after Gage's death, Hanna and Antonio Damasio, using Gage's skull and the iron bar, reconstructed the pathway through his brain (fig. 8.7). They realized that the prefrontal cortex was damaged, particularly the underside, where the ventromedial prefrontal cortex and the orbitofrontal cortex are located—regions that are extremely important for emotion, decision making, and moral behavior.

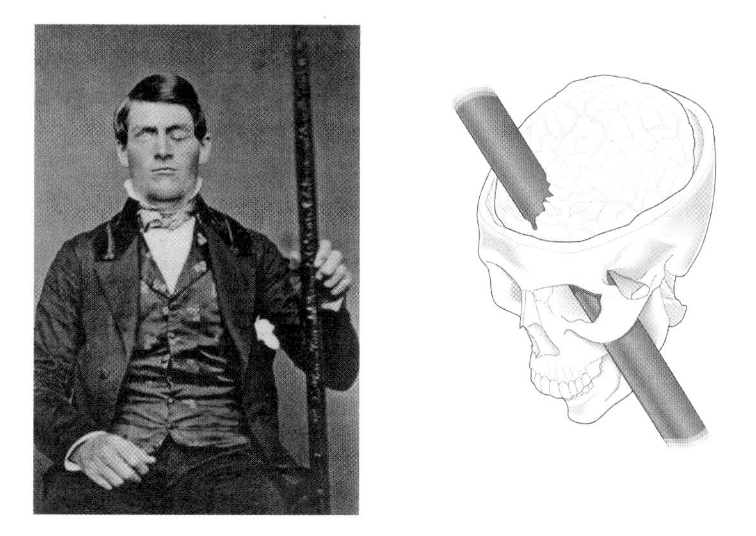

Figure 8.7. Phineas Gage with the iron bar that injured his brain (*left*); reconstruction of the iron bar's pathway through Gage's brain (*right*)

Joshua Greene, an experimental psychologist, neuroscientist, and philosopher at Harvard, has made use of a fascinating puzzle known as the "trolley problem" to study how emotion affects our moral decision making.[12] The trolley problem has numerous variations, but the simplest poses two dilemmas (fig. 8.8). The *switch dilemma* goes like this:

> A runaway trolley whose brakes have failed is approaching a fork in the track at top speed. If you do nothing, the trolley will stay to

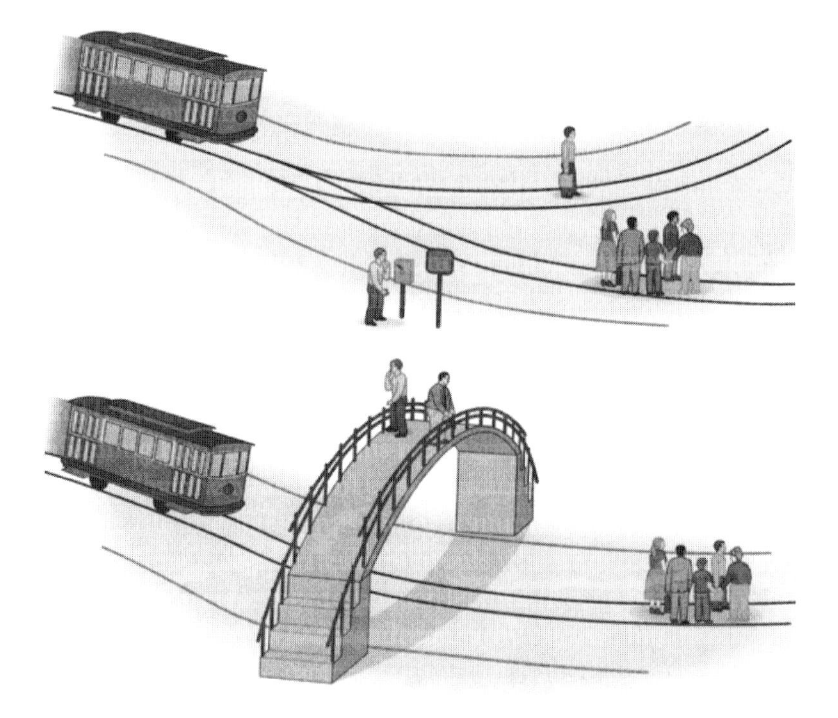

Figure 8.8. The runaway trolley problem: the switch dilemma (*top*) and the footbridge dilemma (*bottom*)

> the right, where it will run over five travelers. All five of them will die. However, if you divert the trolley to the left—by flipping a switch—the trolley will hit and kill one traveler. What do you do? Are you willing to intervene and change the path of the trolley?

Most people agree it is morally permissible to divert the trolley. The decision is based on simple arithmetic: it's better to kill fewer people. Some moral philosophers even argue that it is immoral *not* to divert the trolley, since such passivity leads to the death of four additional people. But what about this scenario, the *footbridge dilemma*?

> You are standing on a footbridge over the trolley track. You see a trolley racing out of control toward five travelers. All five travelers will die unless the trolley can be stopped. Standing next to you on the footbridge is a large man. He is leaning over the railing, watching the trolley hurtle toward the travelers. If you give the

man a push, he will fall over the railing and into the path of the trolley. Because he is so big, he will stop the trolley from killing the travelers. Do you push the man off the footbridge? Or do you allow five travelers to die?

The facts are the same in both scenarios: one person must die in order for five people to live. If our decisions were perfectly rational, we would act identically in both situations. We'd be as willing to push the man as we are to divert the trolley. Yet almost nobody is willing to push another person onto the tracks. Both decisions lead to the same violent outcome, yet most people view one as moral and the other as murder.

Greene argues that pushing the man feels wrong because the killing is direct: we are using our body to hurt his body. He calls this a *personal* moral decision. In contrast, when we switch the trolley onto a different track, we aren't directly hurting someone else. We are just diverting the trolley: the ensuing death seems indirect. In this case, we are making an *impersonal* moral decision.

What makes this thought experiment so interesting is that the fuzzy moral distinction—the difference between personal and impersonal moral decisions—is built into our brain. It doesn't matter what culture we live in or what religion we subscribe to: the two trolley scenarios trigger different patterns of activity in the brain. When Greene asked study participants whether or not they should divert the trolley, their conscious decision-making machinery was turned on. A network of brain regions assessed the various alternatives, sent the verdict onward to the prefrontal cortex, and the people chose the clearly superior option. Their brain quickly realized that it was better to kill one person than five people.

However, when participants were asked whether they would be willing to push a man onto the track, a separate network of brain regions was activated. These regions are associated with the processing of emotions, both for ourselves and for others. People in the study couldn't justify their moral decisions, but their certainty never wavered. Pushing a man off a bridge just *felt* wrong.

Such research reveals the surprising ways in which our moral judgments are shaped by our unconscious emotions. Even though we can't explain these urges—we don't know why our heart is racing or why our stomach feels queasy—we are nevertheless influenced by them. While feelings of fear and stress can lead to aggression, the fear of harming someone else can keep us from engaging in violence.

Studies of other people with brain damage similar to Elliot's and Gage's—that is, damage to the ventromedial prefrontal cortex—suggest that this part of the brain is very important for integrating emotional signals into decision making. If it is, then we might expect these people to make very different kinds of decisions in Greene's trolley problem. They might view it as essentially an accounting question. Five lives for one? Sure, use the oversized man to stop the trolley. In fact, when faced with this dilemma, people with damage to the ventromedial prefrontal cortex are four or five times more likely than ordinary people to say "Push the guy off the footbridge" in the name of the greater good.

This finding underscores the theory that different kinds of moralities are embedded in different systems in the brain. On the one hand we have an emotional system that says, "No, don't do it!" like an alarm bell going off. On the other hand we have a system that says, "We want to save the most lives, so five lives for one sounds like a good deal." In ordinary people these moralities compete, but in people with Gage's kind of brain damage one system is knocked out and the other is intact.

THE BIOLOGY OF PSYCHOPATHIC BEHAVIOR

What about psychopaths, people who would have no difficulty deciding to push somebody off the footbridge? Research on psychopathy indicates that it is primarily an emotional disorder with two defining features: antisocial behavior and lack of empathy for other people. The first can result in horrendous crimes, the second in lack of remorse for those crimes.

Kent Kiehl at the University of New Mexico drives a mobile fMRI machine to prisons to scan the brains of prisoners, many of whom are psychopathic, as indicated by their scores on a standardized checklist. He wants to see whether moral reasoning, or lack of it, can be used to understand the mind of the psychopath—and whether understanding the mind of the psychopath can improve our understanding of moral reasoning.

Greene's theory would predict that psychopaths don't have the emotional response that says pushing the man off the footbridge just feels wrong. They would be likely to go with the numbers, one life for five. But psychopaths are not like people with brain damage; psychopaths work very hard to seem normal, to blend in. To capture what they're really thinking, Kiehl watches not only what the prisoners do but how quickly they do it. For example, a psychopath may be able to hide an emotional reaction to a stimulus—a word or visual image—but he can't do it quickly, and brain imaging will capture his initial reaction.

Using brain imaging, Kiehl found that psychopathic inmates have more gray matter in and around the limbic system than do non-psychopathic inmates or non-inmates. The limbic system, which includes the amygdala and the hippocampus, comprises the regions of the brain involved in how we process emotions. Moreover, the neural circuitry connecting the limbic system to the frontal lobes of the cortex is disrupted in psychopathic inmates. Kiehl notes that several studies have found less activity in those neural circuits when psychopathic prisoners engage in emotional processing and moral decision making.[13]

If psychopathic behavior is based in biology, what does this mean for free will, for individual responsibility? Do these built-in neural processes lead inexorably to certain decisions, or does our conscious sense of morality, our cognitive mental function, have the last word?

This question is becoming increasingly salient in the criminal justice system. Judges look to psychologists and neuroscientists for help in understanding the value and limitations of scientific findings. They want to know if the findings are highly reliable, what they mean in terms of behavior, and how they should be used in a court of law to improve the fairness of the judicial system. The U.S. Supreme Court, for example, recently ruled that a sentence of life in prison without parole for juvenile criminals is unconstitutional. The justices pointed to findings from brain science which indicate that adolescents and adults use different parts of their brain to control behavior.

Most neuroscientists think we should be held responsible for our actions, but the opposing argument has some validity. Should people with brain damage that leaves them incapable of making appropriate moral judgments be treated the same way as people who can make moral judgments? What neuroscience reveals about this question is going to affect our legal system and the rest of our society in the decades to come.

Studies of psychopaths are likely to have a major impact not only on our understanding of how people can be influenced to make appropriate decisions but also on the development of new kinds of diagnoses and new kinds of treatments. Research suggests that both genes and environment contribute to psychopathy, as they do to other disorders. In his continuing search for biomarkers of psychopathy, Kiehl has recently expanded his brain-imaging studies to include young people who show signs of psychopathic traits.[14] This is important because not everyone with psychopathic traits becomes a violent criminal. If scientists could identify children who have a tendency toward psychopathy, they might be able to develop behavioral therapies to head off future violent behavior. If a malfunction of some region of the brain is identified, perhaps some other region of the brain could be encouraged to take over and suppress violent aspects of behavior.

LOOKING AHEAD

Studies of emotion from Darwin and James onward support Damasio's contention that the philosopher René Descartes was in error when he claimed that emotion and reason, body and mind are separate. Fear is a case in point: we cannot simply put mind over matter and reason our way out of post-traumatic stress disorder or chronic anxiety. Studies of how animals learn fear, coupled with imaging studies of the human brain, have given us an understanding of where and how fear operates, including how our brain consolidates the memory of fear. Now innovative psychotherapy and drugs are beginning to help people with anxiety disorders unlearn fear.

Emotion is integral to any personal, social, or moral decision we make. Scientists have found that people with damage to regions of the brain that integrate emotional signals into decision making have great difficulty reaching even simple, everyday decisions. And because they are also unable to engage emotion in moral decision making, these people often make different choices in moral quandaries than people without such brain damage.

Imaging studies have revealed that people who exhibit psychopathic behavior have abnormalities in several areas of the brain concerned with emotional processing and moral functioning. These abnormalities lead to a profound lack of empathy and connection to others. Research in this area is complicated by society's reaction to the crimes committed by the psychopathic prisoners being studied, but if scientists can identify biological and genetic markers of the disorder, treatment and possibly prevention may follow—and along with them a greater understanding of the basic biological mechanisms underlying our moral functioning.

9

THE PLEASURE PRINCIPLE AND FREEDOM OF CHOICE: ADDICTIONS

We have seen that normal fear can spiral into post-traumatic stress disorder, leaving people unable to cope with everyday life. Likewise, our normal attraction to pleasure can go into overdrive, causing the brain to produce an excess of dopamine and resulting in addiction. That addiction may be to substances, such as drugs, alcohol, or tobacco, or to activities, such as gambling, eating, or shopping.

Addiction creates havoc in people's lives. It may cost them their job, their health, or their spouse. They may end up in poverty or in prison. Sometimes, addiction leads to death. People who are addicted do not want to keep doing what they are doing, yet they cannot stop—repeated abuse has eroded the brain's ability to control desires and emotion. Thus, addiction robs us of our will, our ability to select freely among several possible courses of action.

Addiction to substances takes an enormous toll on our society, with an estimated economic cost of over $740 billion annually in the United States. That economic cost climbs far higher if we consider compulsive disorders that are similar to addiction, such as pathological gambling and overeating. The human cost of addiction, for individuals and for society, is incalculable. While we have made progress over the past several decades in treating people with certain kinds of addictions, such as alcoholism, available therapies for most addictions, whether behavioral approaches or medications, have proved inadequate. Fortunately, scientists have made important advances over the last thirty years in understanding the biology of addiction, raising the hope that new treatments will emerge from these new insights.

In the past, addiction was considered to be a manifestation of weak moral character. Today, we understand that it is a mental disorder, a malfunction of the brain's reward system, the neural circuitry responsible for positive emotions and the anticipation of rewards. This chapter introduces us to the brain's reward system and explains how addiction manipulates it. We learn about the stages of addiction and explore various avenues of research. Finally, we learn about new methods of treating people with these chronic disorders.

THE BIOLOGICAL BASIS OF PLEASURE

All of our positive emotions, our feelings of pleasure, can be traced to the neurotransmitter dopamine. Although our brain contains relatively few dopamine-producing neurons, they play an outsized role in the regulation of behavior, largely because of their intimate involvement with the production of pleasure.

First discovered in the 1950s by the Swedish pharmacologist Arvid Carlsson, dopamine is released primarily by neurons in two regions of the brain: the ventral tegmental area and the substantia nigra (fig. 9.1). Neurons in the ventral tegmental region extend their axons to the hippocampus, which is involved in the memory of people, places, and things, and to the three most important brain structures for regulating emotion: the amygdala, which orchestrates emotion; the nucleus accumbens, a region of the striatum that mediates the impact of emotion; and the prefrontal cortex, which imposes will and control on the amygdala. This communications network, known as the mesolimbic pathway, is the major network in the brain's reward system. It puts dopamine-producing neurons in a position to broadcast information widely, including to regions throughout the cerebral cortex.

Soon after Carlsson discovered dopamine, James Olds and Peter Milner, two neuroscientists at McGill University, explored the neurotransmitter's function further.[1] They began by implanting an electrode deep in the center of a rat's brain. The placement of the electrode was largely

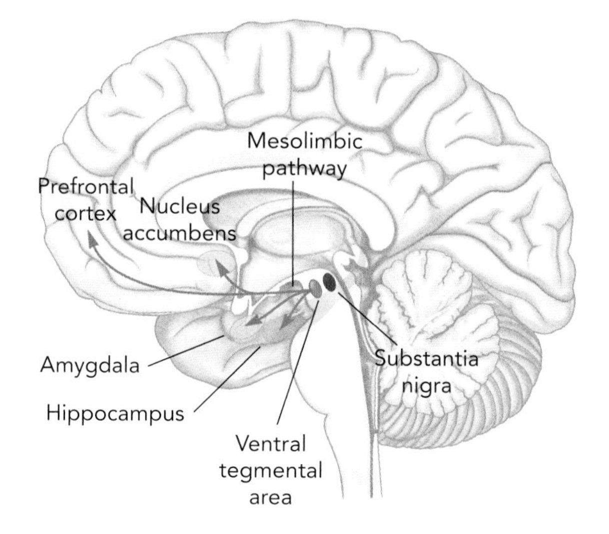

Figure 9.1. The communications network formed by dopamine-producing neurons in the mesolimbic pathway is the key pathway in the brain's reward system.

happenstance, but it turned out that Olds and Milner had inserted it right next to the nucleus accumbens, a crucial component of the mesolimbic pathway (fig. 9.1). They then installed a lever in the cage of the rats that would allow the animals to administer a small jolt of electricity to their brain in the neighborhood of the nucleus accumbens.

The current was so weak that the scientists could not feel it when they applied it to their skin, yet it was pleasurable to the rats when applied to the nucleus accumbens. They would press the lever over and over and over again to produce the desired stimulus. In fact, the pleasure from the electrode was so intense that the animals soon lost interest in everything else. They stopped eating and drinking. They stopped all courtship behavior. They just crouched in the corner of their cage, transfixed by their bliss. Within days, many of the rats died of thirst.

It took several decades of painstaking research before Olds and Milner, and eventually others, discovered that the rats were suffering from an excess of dopamine. Electrical stimulation of the nucleus accumbens had triggered the release of massive amounts of this neurotransmitter, overwhelming the animals with pleasure.

THE BIOLOGY OF ADDICTION

The standard view of a reward is that it is something that makes us feel happy or feel good. Maybe it's chocolate cake, or a new gadget, or a beautiful work of art. Neuroscientists take a slightly different view: a reward is basically any object or event that produces "approach" behavior and leads us to spend attention and energy on it. By reinforcing approach behavior, rewards help us learn.

Specialized regions in the brain appeared early in evolution to regulate our responses to pleasurable stimuli in the environment, such as food, water, sex, and social interactions. All drugs of abuse act on this reward system. Each drug acts on a different target, but in every case the net effect is to increase the amount and persistence of dopamine in the brain. Activation of dopamine signaling, along with activation of several other important reward signals that vary from drug to drug, is responsible for the initial high that people experience on drugs.

Wolfram Schultz, a neuroscientist at the University of Cambridge, has studied the role of rewards in learning.[2] Schultz's experiments with monkeys drew on Pavlov's early experiments with conditioned learning in dogs. Schultz would play a loud tone to monkeys, wait for a few seconds, and then squirt some drops of apple juice into their mouths. While the experiment was unfolding, Schultz monitored the electrical activity inside individual dopamine-producing neurons in the animals' brains. At first, the neurons didn't fire until the juice was delivered. However, once the animals learned that the tone predicted the arrival of juice, the same neurons began firing at the sound of the tone—that is, at the prediction of reward instead of the reward itself. To Schultz, the interesting feature about this dopamine learning system was that it is all about expectation.

The expectation of reward helps us form habits. A good habit, one that is adaptive, helps us survive by enabling us to perform many important behaviors automatically, without thinking about them. Adaptive habits are promoted by the release of dopamine into the prefrontal cortex and the striatum, the areas of the brain involved with control and with reward and motivation. The release of dopamine not only creates a feeling of pleasure, it also conditions us. Conditioning, as we know, creates a long-term memory that enables us to recognize a stimulus the next time we see it and to respond accordingly. If the stimulus is positive, as in the case of adaptive habits, conditioning motivates us to pursue it. For example, if you eat a banana and find it delicious, the next time you see a banana you will feel motivated to eat it.

Addictive drugs, whether legal or illegal—our body doesn't make a distinction—also stimulate dopamine-producing neurons in the brain's reward system. In this case, however, the result is greatly increased dopamine concentrations in the prefrontal cortex and the striatum. The excess dopamine generates intense pleasure and creates a conditioned response to the environmental cues that predict pleasure. Such cues—say, the smell of cigarette smoke or the sight of a needle—elicit an intense craving for the drug, which, in turn, elicits drug-seeking behavior.

Why do some substances, such as cocaine, produce addiction rather than an adaptive habit? Normally, when dopamine binds to receptors on target cells it is taken up and removed from the synapse within a short period of time. However, brain imaging reveals that cocaine, a highly addictive drug, interferes with the removal of dopamine from the synapse. As a result, dopamine lingers there and continues to produce pleasurable feelings that persist beyond those produced by ordinary physiological stimuli. In this way cocaine hijacks the brain's reward system.

This hijacking takes place in several well-defined stages, beginning with the addictive process itself, in which a drug takes over the brain's reward system, and ending with an inability to resist taking the drug. Every drug of abuse that we are aware of increases concentrations of dopamine in the pleasure centers of the cortex, and this increased dopamine is believed to produce the rewarding effects that define the drug experience. Many addictive drugs release additional chemicals that mediate reward.

As a person continues to take the drug, however, he or she builds up a tolerance to it. The dopamine receptors no longer respond as effectively as they did before. The same amount of the drug that initially produced a high—the pleasurable feeling—now produces a normal feeling. As a result, the person needs more of the drug to produce an equivalent high. Nora Volkow, director of the National Institute on Drug Abuse and a pioneer in the study of how addiction affects the human brain, has documented this process in a series of imaging studies showing that the striatum stops responding once a person has used cocaine for some time.[3]

At first glance, drug tolerance doesn't seem to make sense. If a person takes a drug to feel good but that drug is not effective at increasing dopamine (which causes the pleasurable feeling), then what is the point of taking the drug? This is where positive associations come into play. An addicted person has learned to associate the drug with a certain place, certain people, certain music, and a certain time of day. Paradoxically, these associations rather than the drug itself often lead to the most tragic aspect of addiction: relapse.

Relapse is possible even after a person has given up drugs for weeks, months, or even years. The memory of the pleasurable drug experience and the cues associated with it essentially persist forever. Exposure to those cues—the sight or smell of the drug, walking down a street where the person used to buy the drug, or bumping into people who used the drug—triggers a tremendous urge to use the drug again.

A particularly interesting study of addiction by Lee Robins, a sociologist at Washington University in St. Louis, involves Vietnam veterans who had become hooked on very high quality heroin while overseas. Amazingly, most of them were able to conquer their addiction when they returned to the United States because none of the cues that had encouraged them to use heroin in Vietnam were present at home.[4]

RESEARCH ON ADDICTION

Because of the ease with which addicted people relapse, we now know that addiction is a chronic disease, like diabetes. People can be helped to avoid relapse, but recovery is a lifelong process requiring great effort and vigilance on the part of the addicted person. To date, there is no cure for addiction, but in recent years scientists have made progress in understanding the disorder.

The first important avenue of investigation is brain imaging, as pioneered by Volkow. Imaging enables us to look inside the brain of an addicted person and see what areas are disrupted. These abnormal patterns of activity help to explain why some people cannot control the urge to take drugs, even though the drugs themselves are no longer pleasurable.[5]

In one study, Volkow gave cocaine to addicted people and to people who were not addicted and then compared images of their brains, using positron emission tomography (PET). She expected to see a lot of activity in the main reward areas of the brain, and that's exactly what she did see—in the brains of people who were not addicted. As dopamine concentrations increased, activity in their reward system spiked dramatically. To her surprise, however, she saw almost no activity in the brains of addicted people. These findings explain how our brain builds up a tolerance to drugs.[6]

Volkow was drawn to the study of addiction because of the insights it offers into the normal workings of the brain. As she has pointed out in a personal communication, she has always been interested in understanding how the human brain controls and sustains its behavior.

The study of drugs of abuse and addiction enabled her to investigate a condition in which the capacity to control oneself is disrupted. Brain imaging, in turn, enabled her to carry out studies in human beings afflicted by addiction. By studying the effects of drugs in the brain, she was able to gain insights into the neural circuits that shape behavior in response to environmental contexts and exposures and how these are subjectively experienced by the individual. In particular, she was interested in changes that are associated with pleasure, fear, and cravings.

Similarly, by studying the brain of a person who is addicted and comparing it with the brain of someone who is not, she could identify the neural circuits that are disrupted and explore how this disruption relates to the disruption of self-control. From these studies, it became clear that addiction is a disease of the brain and that the changes triggered by drug exposure influence circuits in the brain that process motivation and reward.

The second avenue of research into addiction, as Darwin might have predicted, involves experiments on animals. Because the dopamine system exists in similar form in many other animals, scientists can study craving and addiction in monkeys, rats, and even flies. While many advances in modern medicine have come about through the use of animal models, that is especially true in addiction.

Animals readily become addicted to drugs, and the physiological and anatomical changes in their brains are similar to those in people. Addicted animals no longer show activity in the reward areas of the brain. Furthermore, the same factors that increase the likelihood of addiction in people also increase the likelihood of addiction in animals. We know, for example, that chronic stress will increase vulnerability to drug abuse in rats and in people because such drugs can transiently relieve some of the physiological and emotional consequences of stress. We also know that rats will choose to self-administer and addict themselves to the same range of drugs as people. Moreover, animals given unlimited access to a very potent drug like cocaine or heroin will overdose and kill themselves.

We have also learned from animal models how repeated exposure to a drug of abuse changes the brain's reward system. Some of the changes occur within the neurons that produce dopamine, impairing their function and their ability to send dopamine signals to other regions of the brain. These changes are linked to drug tolerance—the reduced reward that individuals obtain from drugs as they take them repeatedly—as well as to the diminished responsiveness to rewards that people experience during withdrawal (fig. 9.2).

Eric Nestler, at Icahn School of Medicine at Mount Sinai Hospital in New York City, notes that this diminished responsiveness is similar to the inability of people with depression to experience pleasure. In studies of mice addicted to cocaine, Nestler and his colleagues found that "by manipulating the reward pathway in these mice, we were not only able to prevent the rewarding effects of cocaine, but surprisingly, we could push these animals to a point where they were anhedonic—unable to experience pleasure." Nestler has since studied the role of the brain's reward system in depression as well as in addiction.[7]

Scientists have identified the numerous chemical changes in animals' brains that are induced by addictive drugs. Some of the changes are related to a drug's ability to diminish the reward system's sensitivity to dopamine. Other changes are related to a drug's ability to promote compulsive, repetitive behavior. For example, scientists have found a molecule that modifies the expression of certain genes in a way that helps perpetuate memory. By disrupting the activity of this molecule in rats that are addicted to morphine, the scientists could eliminate the animals'

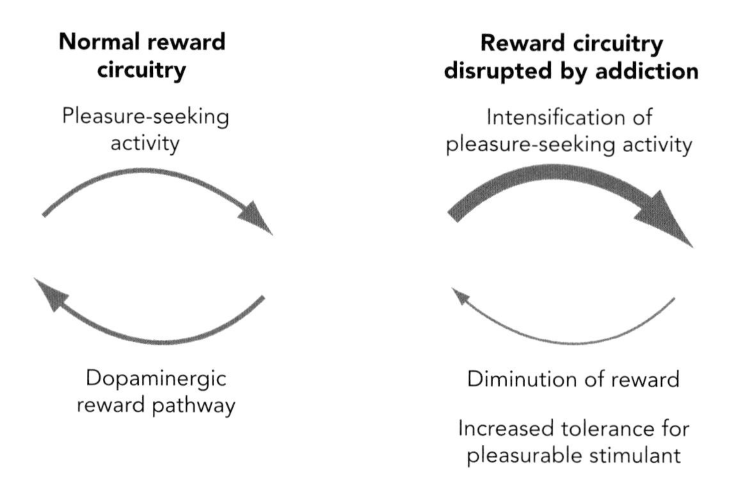

Figure 9.2. The brain's normal reward circuitry is disrupted by addiction.

cravings for the drug.[8] Such research raises the intriguing possibility that future treatments for addiction will focus not just on the pleasure pathway, but on our memory of pleasure as well.

Other drug-induced changes in the animals' brains bring about positive associations between the drug experience and environmental cues. Both kinds of changes help drive addiction. Thus, although an animal taking the drug will build up tolerance to the drug itself, the addiction continues because craving is triggered by cues present in the environment. Increasingly, advanced brain-imaging techniques and postmortem examinations of the brains of human addicts have confirmed that findings from animal models apply to people as well.

Perhaps the most surprising finding to emerge from animal models is that the heritability of addiction is moderately high: roughly 50 percent. This means that the genetic risk of addiction is greater than that of type II diabetes or high blood pressure.[9] The remaining 50 percent results from the interaction of environmental factors and genes. "Ultimately, the ability of environmental stimuli to influence an organism requires changes in gene expression," says Nestler, who has explored the ways in which drug addiction alters gene expression.[10] Scientists are just now developing molecular genetic techniques that will enable us to pinpoint the genes involved in addiction.

Nestler has found several genes in the reward systems of animals that, when modified, dramatically reduce the vulnerability to addiction.[11] Identifying the specific genes that confer the risk of addiction and understanding how environment interacts with those genes will guide the development of better diagnostic tests and treatments.

The third kind of research into addiction is the epidemiological study, which tracks the incidence or prevalence of a particular addiction in a particular population over a particular period of time. Thanks to epidemiological studies, we now know that the use of certain addictive drugs increases the likelihood of using other addictive drugs.

Denise Kandel of Columbia University has been instrumental in uncovering some of those links. She has used epidemiological studies of young people to show that smoking is a powerful first step toward cocaine or heroin addiction.[12] That finding raised the question of whether young people start with nicotine because it is the first drug available, or whether nicotine does something to the brain that makes it more vulnerable to other substances and to addiction.

Kandel, Amir Levine, and their colleagues studied this question in mice and found that exposing the animals to nicotine modifies their dopamine-receiving neurons in such a way that they respond more powerfully to cocaine. In contrast, giving the animals cocaine first has no effect on their subsequent response to nicotine.[13] Thus, nicotine primes the brain for cocaine addiction.

Society has taken great pains to discourage people from smoking, and it is quite likely that reducing the number of smokers will reduce other sorts of addictions as well.

OTHER ADDICTIVE DISORDERS

Some compulsive disorders—those involving eating, gambling, and sexual behavior—are very similar to drug addiction. We know that addiction is an exaggerated response to a given reward, and it is likely that the same parts of the brain activated by addictive substances are also activated by food, money, and sex. Studies that compared brain images of drug-addicted people and obese people have found similar changes in the brain. Just as addicted people often show reduced activity in parts of the reward system when taking drugs—they have become conditioned to the pleasure—so obese individuals show reduced pleasure when eating. Research has found that the reward system of obese people tends to be less responsive to dopamine and to have a lower density of dopamine receptors.

Kyle Burger and Eric Stice at the Oregon Research Institute conducted an interesting study of adolescents' eating habits.[14] The researchers began by asking 151 teenagers of varying weights about their eating habits and food cravings. Afterward, they put the teens in a brain scanner and showed them a picture of a milkshake, followed by a few sips of the real thing. The researchers then compared the activity of the teens' reward system to their answers to questions about their eating habits.

The teenagers who reported eating the most ice cream showed the least activation of their reward system when consuming the milkshake. This suggests that they ate more in order to compensate for the reduced pleasure they actually got from eating. They had to consume larger quantities (and additional calories) to achieve an equivalent reward, just as someone addicted to drugs would. This finding indicates that obesity results from reward-related changes in the brain, not from gluttony or self-indulgence. Thus, an understanding of the biology of obesity is essential to stopping the stigmatizing of obese people.

Research has shown that obesity has a social component as well: that is, it seems to spread from person to person. Nicholas Christakis of Harvard University and James Fowler of the University of California, San Diego, recently sifted through the handwritten records of 5,124 men and women from the Framingham Heart Study, an ongoing project that was begun in 1948 and that has revealed many of the risk factors associated with cardiovascular disease. The original Framingham researchers had kept careful notes not only on each participant's family members but also on their close friends and colleagues. Because two-thirds of all Framingham adults had participated in the first phase of the study, and their children and grandchildren had participated in subsequent phases, almost the entire social network of the community had been chronicled. Christakis and Fowler constructed a detailed network of personal associations from these records, enabling them, for the first time, to see how a social network influences behavior.[15]

The first variable Christakis and Fowler analyzed was obesity, and here they made a remarkable discovery: obesity seemed to spread through a social network like a virus. In fact, if one person became obese, the likelihood that a friend would follow suit increased by 171 percent. Christakis and Fowler went on to find that smoking also spreads from person to person. When a friend begins smoking, your chances of lighting up increase by 36 percent. Similar percentages apply to drinking alcohol, to happiness, and even to feelings of loneliness.

Studies of the biological and social factors underlying obesity may not only help scientists develop ways to prevent obesity, they may also provide insights into the development of drugs for other kinds of addiction. Self-control will never be easy. But perhaps we can help people with a malfunctioning reward system by making it slightly less difficult to achieve self-control.

TREATING PEOPLE WITH AN ADDICTION

Animal models and other studies have taught us a great deal about how to treat people with an addiction. First of all, the studies show that addiction is a chronic disease. The notion that a person can go to rehab for a month and be cured is not correct. It's magical thinking.

Second, addiction affects several regions of the brain, several neural circuits. This calls for a multipronged approach to treatment and raises several questions. Can an addicted person's self-control be strengthened through behavioral therapy that helps rein in self-destructive behavior or by medication that improves the functioning of the prefrontal cortex? Can behavioral interventions or medication weaken conditioning, so that when a person sees stimuli associated with the addictive substance he or she doesn't respond to them? Can the reward system be made to respond to natural stimuli, so that things other than drugs will motivate the addicted person?

The most successful treatments for addiction to date are behavioral and involve regimented twelve-step programs like Alcoholics Anonymous. But most addicted people return to using drugs even after completing the best available programs. These high relapse rates reflect the long-lasting changes that take place in the brain during addiction. As we have seen, drug addiction is a form of long-term memory. The brain becomes conditioned to associate certain environmental cues with pleasure, and encountering those cues can trigger an urge to use the drug. The memory of pleasure persists long after an addicted person has stopped taking a drug; that is why maintaining treatment—even after repeated relapses—is so important.

The goal of medications would be to help an addicted person forget the pleasure associated with an addictive drug and counteract the powerful biological forces that drive addiction, thereby enhancing the effectiveness of rehabilitation and psychosocial treatment. We have seen that behavioral therapies and medication both work through biological processes in the brain and that they are frequently synergistic. One of the central challenges in treating addiction is to translate our increasing knowledge of the brain's reward circuits into new therapies.

Unfortunately, pharmaceutical companies have devoted very little effort to developing drugs to treat addiction. One reason is their perception that they cannot recover their research costs from addicted people. Nevertheless, basic research has led to some important medications that reduce cravings.

Nicotine-replacement drugs, for example, target the same areas of the brain as nicotine itself, but they do so in a way that helps reduce the craving for cigarettes. Methadone binds to the same receptors that are activated by heroin, but it stays on the receptor for a very long period of time, thus reducing the intensity of the emotional response. Although methadone itself is an addictive drug, addiction to methadone does not disrupt day-to-day behavior as severely as addiction to heroin does. In addition, methadone is a prescription drug that is available legally, whereas heroin is an illegal drug that must be bought on the black market, often under risky circumstances.

Current treatments for addiction are deeply flawed, but, as we have seen, brain-imaging studies, animal models of addiction, and epidemiological studies are all contributing to an increased understanding of the changes in the brain's reward system that underlie addiction. Many scientists are working on treatments aimed at restoring normal activity in the dopamine-producing circuits of the brain, through medication, behavioral therapy, and genetic therapy. Eventually, this research into treatment may enable us to develop ways of preventing addiction.

LOOKING AHEAD

The health care system has for the most part removed itself from the screening and treatment of people addicted to drugs because addiction is widely believed to be a behavior of choice—a bad behavior by a bad person. This belief stigmatizes addicted people.

The question of exercising our will in the context of addiction is a difficult one, because drugs target the parts of our brain that control our ability to make decisions. As we have seen, addiction is a complex interplay between conscious and unconscious mental processes. It starts with a conscious decision to obtain drugs, but the drugs stimulate neurons to produce dopamine, and sometimes other chemicals, in the brain. Eventually, this unconscious activity, and the changes it causes in brain function, takes over. While an addicted person may have made the initial choice to experiment with the drug, the subsequent brain disorder diminishes his or her ability to choose freely.

Education and science are our best means of eliminating stigma and thereby enabling individuals and society to behave in a more rational manner toward addicted people. Drug overdoses are now estimated to be the leading cause of death among Americans under the age of fifty.[16] Studies have found that 40 percent of eighteen- to nineteen-year-olds in the United States have been exposed at least once to an illegal drug, and 75 percent or more have been exposed to alcohol. Some of them—approximately 10 percent—will become addicted; the others will not. Given that the risk of addiction is strongly shaped by genetics, it's important that we approach addiction as a brain disorder, not as a moral failing, and that we provide treatment, not punishment, for people with addictions.

10

SEXUAL DIFFERENTIATION OF THE BRAIN AND GENDER IDENTITY

Most of us have a strong sense of gender identity—of being a boy or a girl—early in life. Consequently, we grow up behaving in ways that are more or less typical of other boys or girls in our society. Usually, our gender identity conforms to our anatomical sex, our genitals and reproductive organs, but not always. We may have a male body but feel like a girl or a woman, or have a female body but feel like a boy or a man. This variance is possible because our sex and our gender identity are determined separately, at different times in the course of development.

Gender identity is our sense of where we belong on the continuum of sexuality, of being a man, a woman, or neither, or both. It encompasses our biological development, our feelings, and our behavior. So while gender identity may vary widely among individuals, it is a function of the normal sexual differentiation of the brain. It is because we can learn so much about ourselves from the study of gender identity that I digress from consideration of brain disorders to include this chapter on sexual differentiation of the brain.

For people whose gender identity is at odds with their anatomical sex—that is, for people who are transgender—the feeling of being in the wrong body begins in childhood and may intensify in adolescence and adulthood. The tension between their outward appearance—which creates a constellation of social expectations regarding behavior—and their inner feelings causes confusion and distress and may make interactions with others difficult. As a result, transgender people may experience anxiety, depression, or other disorders. Moreover, transgender people often face severe discrimination and physical danger.

Gender identity is not the same thing as sexual orientation, a person's romantic attraction to the opposite sex, the same sex, or both sexes. At present, we know too little about the biology of sexual orientation to discuss it here.

Where does our sense of gender identity come from? Is it determined before birth, or is it a social construct? In this chapter we first consider sexual differentiation, the genetic, hormonal, and structural changes that take place during development and that determine our anatomical sex. Next, we look at gender-specific behavior. We explore what differences between male and female behavior tell us about physical differences between the male and female brain. We then learn about genes that can cause gender identity and anatomical sex to diverge. Together, these findings are beginning to give us a much more nuanced picture of human gender identity and how it is influenced by the brain.

We learn from a gifted scientist how he felt growing up as a boy in a girl's body and later transitioning from a woman to a man. Finally, we look at some of the questions surrounding how best to support children and adolescents whose gender identity is different from their sex at birth.

ANATOMICAL SEX

The word "sex" is used in three ways to describe the biological differences between men and women. *Anatomical sex*, as we have seen, refers to overt differences, including differences in the external genitalia and other sexual characteristics, such as distribution of body hair. *Gonadal sex* refers to the presence of male or female gonads, the testes or ovaries. *Chromosomal sex* refers to the distribution of the sex chromosomes between women and men.

Our DNA is distributed into twenty-three pairs of chromosomes (fig. 10.1). Each pair is made up of one chromosome from our mother and one from our father. The chromosomes in any given pair between 1 and 22 have a similar, but not identical, DNA sequence.

The two chromosomes in the twenty-third pair—the X and Y

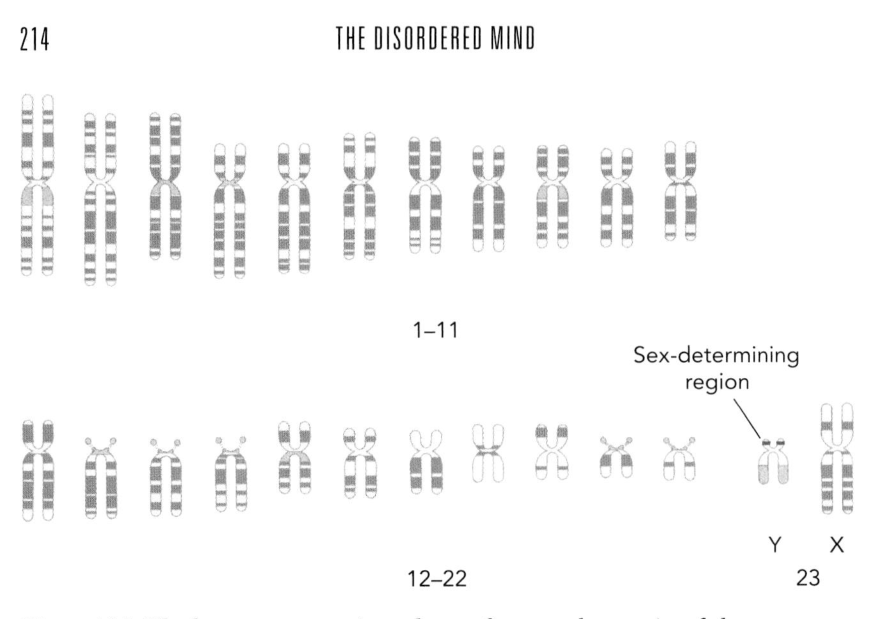

Figure 10.1. The human genome is made up of twenty-three pairs of chromosomes; the twenty-third pair determines anatomical sex.

chromosomes—are very different from each other. These are the chromosomes that determine our anatomical sex. The X chromosome, the female chromosome, is roughly the same size as the other forty-four chromosomes; the Y chromosome, the male chromosome, is considerably smaller. Women have two copies of the X chromosome, so they are genetically XX; men have one copy of X and one of Y, so they are genetically XY.

How does the Y chromosome produce a boy? Initially, every embryo has an undifferentiated gonadal precursor called the *genital ridge*. Around the sixth or seventh week of gestation, a gene on the Y chromosome called *SRY* (sex-determining region Y) initiates the process of becoming male by directing the undifferentiated genital ridge to develop into the testis (figs. 10.2 and 10.3). Once the testis has developed, the embryo's sexual fate is sealed further by the action of the hormones released by the testes, such as testosterone. By about the eighth week of gestation, the testes of the male fetus release almost as much testosterone as those of a boy at puberty or an adult man. That massive release of testosterone is responsible for almost every aspect of being male, including body form and brain characteristics.

At about six weeks of gestation, an embryo with two X genes begins the process of female sexual development: ovaries develop, and the sexual differentiation of the body and aspects of brain development follow the female pathway (figs. 10.2 and 10.3). The embryo does not require a massive release of hormones from the ovaries in order to become female.

GENDER-SPECIFIC BEHAVIOR

Male and female animals display clear differences in their sexual and social behaviors. Indeed, in every species, including our own, each individual exhibits a set of behaviors that is typical of its sex: biological males behave in a manner that is typical of males, and biological females behave in a manner that is typical of females.

Gender-specific behaviors, particularly sexual and aggressive behaviors, are remarkably similar across species, which indicates that such behaviors have been carefully conserved over the course of evolution. This suggests, in turn, that the neural circuits driving the behaviors are also very similar and highly conserved. The signals that trigger gender-specific behaviors, however, are usually specific to a given species.

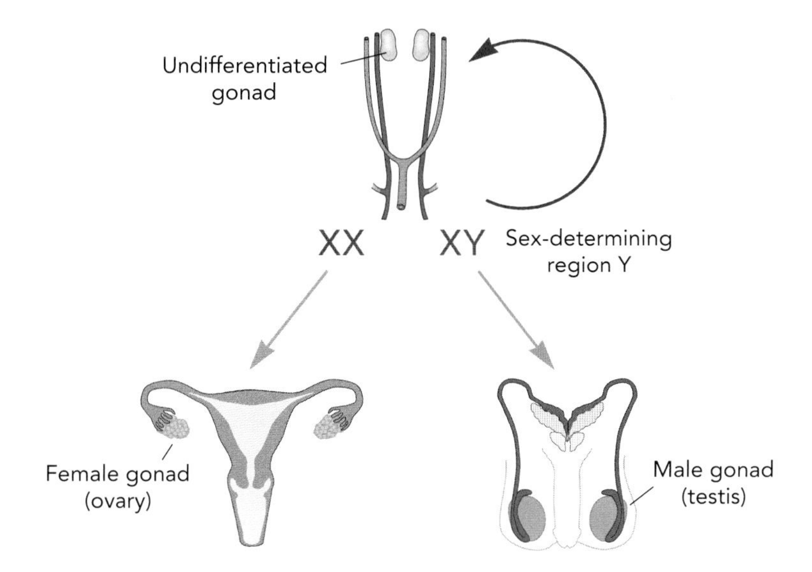

Figure 10.2. The differentiation of an embryo into male or female occurs in the sixth or seventh week of gestation.

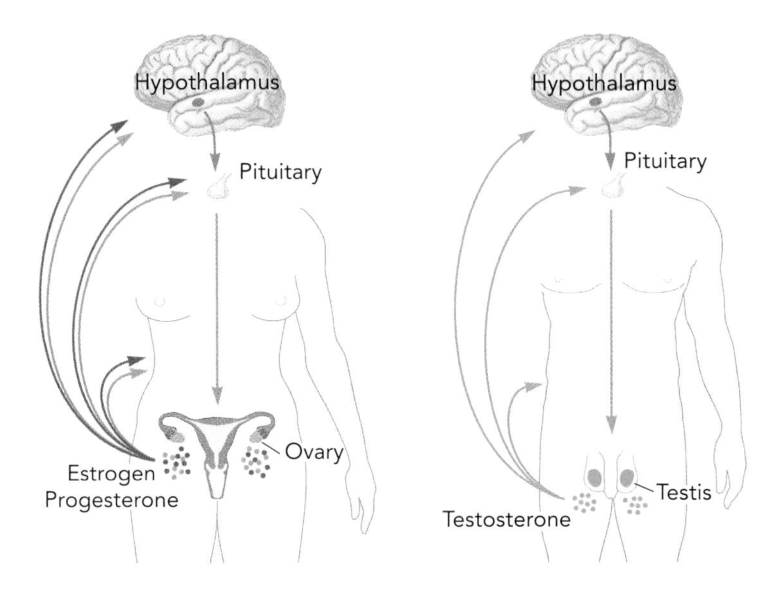

Figure 10.3. The release of male or female hormones gives rise to the male or female body form and brain characteristics.

In the American flicker, for example, only one signal triggers gender-specific behavior: a black pattern on the face of the male bird that looks like a mustache. If a male flicker sees another flicker with a mustache, he will attack the other flicker because he assumes it is a male. If you were to paint a mustache on the face of a female flicker, the male would attack her, and if you masked the mustache of a male flicker, the other males, assuming he was a female, would attempt to mate with him. Similarly, gender-specific behavior in mice is triggered by olfactory cues, called *pheromones*, that are emitted by other male or female mice, and humans are particularly sensitive to visual and auditory cues, a fact successfully exploited by the pornography industry.

Once we know what signals trigger gender-specific behaviors, we can study how the brain controls the display of those behaviors. Norman Spack of the Gender Management Service at the Boston Children's Hospital of Harvard Medical School has found that our body releases sex-specific hormones shortly after birth as well as at puberty.[1] These hormones are critical for molding the brain in a gender-specific manner. In boys, the spikes of testosterone are essential for the proper development of neural circuits controlling male-specific behaviors, particularly aggression. Conversely, the release of estrogen in girls primes mating behavior. Without this early release of estrogen, a different set of sex-specific behavior circuits develops, one affecting in particular female-male mating and maternal behavior.

Because mice display clear gender-specific behaviors, Catherine Dulac at Harvard and David Anderson at the California Institute of Technology can use modern genetic and molecular tools to study the mechanisms in the brain that control those behaviors. Their studies have turned up several interesting things about the mouse brain that may translate to the human brain.[2]

First, the neural circuits that control the gender-specific behavior of each sex are present in both sexes. Thus, regardless of a mouse's sex, its brain contains the neural circuitry for both male and female behavior. These circuits are regulated by pheromones, the hormone-like substances released into the environment by other mice. Normally, when a mouse's brain detects a pheromone, it activates the behavior called for by the mouse's sex and represses behavior appropriate to the other sex. Thus, in a female mouse, female-specific sexual or parental behavior would be activated and male-specific behavior would be repressed, and vice versa in a male mouse. Genetic experiments have shown, however, that under certain circumstances male and female mice can display behaviors linked to the opposite sex. A female mouse with a mutant pheromone-detection gene behaves like a male mouse, seeking out female partners, and a male mouse with a mutant pheromone-detection gene behaves like a female mouse, caring for infant mice rather than killing them, as a male normally would.

Second, since the brains of male and female mice are largely similar, their behavior is not determined exclusively by their biological sex. This is important because animals occasionally need to display the behavior of the other sex. Males are paternal during a brief period after mating and the birth of offspring, and females in many species show mounting behavior as a display of dominance.

This bisexual nature of the brain has been observed in fish and reptiles as well as mice and other mammals, and it is believed to be of great importance to the control of gender identity in humans.

SEXUAL DIMORPHISM IN THE HUMAN BRAIN

Do the structural differences in the brain that control gender-specific behavior in male and female mammals also exist in our brain? Advances in high-resolution magnetic resonance imaging (MRI) and genetic technology have revealed that although the male and female brain share many features, there are nonetheless sex-specific structural and molecular differences, or *sexual dimorphisms,* in several regions of our brain. These differences occur in areas that are involved in sexual and reproductive behaviors, such as the hypothalamus, but they also occur in neural circuits linked to memory, emotion, and stress.

So the answer is that yes, clear sexual dimorphisms do exist in the human brain. What we don't know yet is just how those dimorphisms relate to behavior.

In some cases, the relation appears to be fairly straightforward. Scientists think, for example, that the neural circuits responsible for penile erection in male mice and for lactation in female mice translate readily to humans, but there is no agreement on what animal studies can tell us about human behavior beyond that. We have a poor understanding of how sexual dimorphisms in the human brain govern cognitive functions such as gender identity. Moreover, we have made little headway in tracing differences in men's and women's cognitive function to structural differences in the brain.

Progress in this area has been hampered, in part, by controversy over whether cognitive differences between men and woman even exist. Some people argue that sex-specific differences stem from family and societal expectations. Others argue that the differences have a biological basis. If cognitive differences do exist, they are small and represent differences between the means of highly variable male and female populations. In other words, scientists have found far greater variation within each sex than between the sexes.

The existence of some physical differences between the brain of a man and the brain of a woman implies that some of the neural circuitry of the brain is also different, and that sometimes those differences are directly related to differences in behavior. At other times, however, sex-specific behavior appears to result from different ways of activating the same basic circuits. So the question is this: Does our brain contain neural circuits for both male and female behavior, like the mouse brain, or does it have separate neural circuits for men and for women?

New insights into the link between sexual dimorphism in the human brain and gender identity have come from genetic studies. These studies show that some single-gene mutations cause anatomical sex to become disassociated from gonadal and chromosomal sex. For example, anatomical girls with the gene for congenital adrenal hyperplasia (CAH) are exposed to an excess of testosterone during fetal life. The condition is generally diagnosed at birth and corrected, but the girls' early exposure to testosterone is correlated with subsequent changes in their gender-related behavior. The average girl with CAH tends to prefer toys and games typical of boys her same age. A small, but statistically significant, increase in the incidence of homosexual and bisexual orientation also occurs in women who were treated for CAH as children. Moreover, a significant proportion of these women also express the desire to live as men, consistent with their gender identity.

These findings suggest that the sex hormones released in our body before birth influence our gender-specific behavior independently of our chromosomal and anatomical sex. Dick Swaab and Alicia Garcia-Falgueras at the Netherlands Institute for Neuroscience explain why. They note that gender identity and sexual orientation "are programmed into our brain structures when we are still in the womb. However, since sexual differentiation of the genitals takes place in the first two months of pregnancy and sexual differentiation of the brain starts in the second half of pregnancy, these two processes can be influenced independently, which may result in transsexuality."[3]

Similarly, two genetic conditions affecting boys—complete androgen insensitivity syndrome (CAIS) and 5-alpha reductase 2 deficiency—often result in feminized external genitalia. Boys with one of these conditions are mistakenly raised as girls until puberty, but at that point their pathways diverge. The symptoms of 5-alpha reductase 2 deficiency arise from a defect in testosterone processing, not testosterone production, and are largely confined to the developing external genitalia. At puberty, the massive increase in circulating testosterone causes boys with this condition to develop male characteristics: male distribution of body hair, musculature, and, most dramatically, male external genitalia. At this stage, many adolescents choose to adopt the male gender. By contrast, CAIS arises from a defect in androgen receptors throughout the body. Young people with this condition seek medical advice when they fail to menstruate at puberty. Consistent with their feminized appearance, most of them have a female gender identity and a sexual preference for men. They may request surgical removal of the testes and receive supplemental female hormones.

GENDER IDENTITY

Gender identity, as we have seen, begins to be apparent early in childhood and is not based on anatomical sex. That is why even as a child, a person can feel that he or she is trapped in the wrong body, expected to behave one way but feeling and wanting to behave differently. Often, transgender people change their sex—socially, hormonally, surgically, or in each of these ways—to match their gender identity more closely. We see this in the life stories of Ben Barres (fig. 10.4), who grew up transgender and eventually decided to transition surgically from a woman to a man, and Bruce Jenner, who transitioned from a man to a woman.

Ben was born in 1955 as Barbara Barres and changed sex from female

Figure 10.4. Barbara / Ben Barres

to male in 1997. He was an extraordinarily gifted brain scientist and was chair of the neurobiology department at Stanford University from 2008 to 2017. In 2013 he became the first openly transsexual scientist invited to join the National Academy of Sciences.

It is therefore not surprising that when Deborah Rudacille wrote her now classic book on anatomical sex and gender identity in 2006—*The Riddle of Gender*—she introduced a conversation with Barres in her first chapter.

> As early as I can remember, I thought that I was a boy. I wanted to play with boys' toys, play with my brother and my brother's friends and not my sister. I was always being given girls' toys, like Barbie. . . . I wanted to be in the Cub Scouts so bad, and Boy Scouts. Instead I was in the Brownies, and I hated that. We were baking cookies, and I wanted to go camping. . . .
>
> I was remembering just the other day . . . the Girl Scout leader yelling at me, saying, "Why do you always have to be different, Barbara? Why do you always have to be different?" And she was absolutely at her wits' end. [I was] shocked by this because I was always the good kid. You know, I always got good grades and I never got in trouble. I wasn't trying to cause any trouble. . . . And then, because she shocked me so much, I started thinking about it and kind of said to myself, "You know, I guess I am doing something kind of different than the other girls."[4]

After he reached puberty and developed breasts, which he did his best to hide with loose clothes "so that they wouldn't show," Barres experienced increasingly acute discomfort:

> I had this feeling of just being wrong in my body. I just started to feel very uncomfortable and in fact became uncomfortable for the rest of my life, because you have to wear dresses. If you are a doctor, you have to wear a dress to go to the clinic. You have to wear a dress to funerals and weddings. Having to go to my sister's wedding and wear this flowery dress. These are amongst the big traumatic experiences of my life!

> And that sort of discomfort (because I've only changed my sex over the past few years) has characterized most of my life. Just this very, very uncomfortable feeling about being female—every aspect of it. But I didn't understand it and I was always very confused about it.[5]

While he was in college, Barres was diagnosed with mullerian agenesis, a congenital condition that resulted in his having ovaries but no vagina or uterus. Young women with this condition usually identify as women and may choose to undergo a medical procedure to create a vagina. For Barres, who had never felt like a girl, the situation was different:

> I remember talking to these doctors and they were saying that they were going to construct an artificial vagina, and I never had any say in the matter. They never asked me if I wanted it. . . . They would come in and they would go out, but they would never ask me how I felt. And I had feelings! I felt very confused about the whole thing, like why are they going to do this, and I really don't feel female, and I didn't think that I particularly wanted a vagina. But on the other hand, I was a girl and I should have a vagina. It didn't seem like there was any choice really. . . .[6]

Barres graduated from the Massachusetts Institute of Technology and went on to medical school at Dartmouth. He obtained a Ph.D. in neurobiology from Harvard and joined the faculty at Stanford University in 1993. In 1997, he made the difficult decision to undergo a sex change operation. Barres explains how it came about:

> Here I am, a doctor. I've been confused about my gender my whole life. . . . And then [I] read this article [about James Green, a well-known transsexual man and activist] and it's like in your face. It was so moving. It was like everything he said was the story of my life. And in the article it mentioned this clinic right down the street . . . so I just contacted them . . . and the next thing you know they were seeing me and saying, "You are a classic case. Would you like to change your sex?" . . .
>
> There was a period of a few weeks where I was pretty stressed

> because I was thinking, "Do I really want to do this?" . . . I never feel like I really do a good job of explaining what it was like, but I didn't sleep a lot of nights, I was suicidal. . . . [I]t's like [my life] was split into two parts. The personal part, which has been very uncomfortable, and the professional part that's been a pleasure. . . .
>
> So, at the time I went to the clinic, I just felt like it was either this or suicide. I didn't see any other alternatives. And it all happened very quickly. Within a few months of being seen, I was on hormones and then within a few months after that, I had my ovaries taken out. . . .[7]

As Barres said later, "I thought that I had to decide between identity and career. I changed sex thinking my career might be over. . . . Very fortunately, my academic colleagues have been incredibly supportive and my fears were far worse than reality."[8] Barres told Rudacille, "I feel like I had this gender issue, I dealt with it, and it's resolved. The most important thing is that I've been happy. I've been so much happier. I enjoy life now."

When asked whether he thinks gender identity is mental or physical, biological or social, Barres replied:

> I think that there is something bimodal about gender. Biologically bimodal, because it's important for evolution and all species have it. Males and females are designed differently, and it's all under the influence of hormone-driven programs, and if you look at behavior, male and female behavior is different, and I don't think that's all social. In fact, some of the best evidence for that comes from transsexuals. If you look at female-to-male transsexuals and the results of their spatial tests before and after testosterone . . . you find that female-to-male transsexuals become more malelike in their spatial abilities after testosterone. So there clearly are some gender-specific things that are controlled by hormones.
>
> . . . [B]ut of course in any spectrum there's going to be something in between. I just think that's biology; it's just the way we are. I would think that a lot of transsexuals feel this way because otherwise why do they feel so strongly from the time they are born that there's something wrong? Why can't they just get

> used to the way they are? That doesn't come from the way society treated me. That comes from deep within.[9]

Bruce Jenner followed a different path, transitioning from a muscular, athletic man to a woman. Jenner was a superb football player in college, but he developed a serious knee injury, requiring surgery that prevented him from returning to the game. Jenner was convinced by L. D. Weldon, who coached Olympic decathlon athletes, to take up the decathlon, a series of ten different track-and-field events.

Under Weldon's training Jenner went on to win the decathlon gold medal at the 1976 Montreal Summer Olympics. Because the decathlon requires so many different skills, the winner of the gold medal is unofficially called the World's Greatest Athlete. Not only did Jenner win, he broke the existing record in the decathlon. He went on to become a broadcaster for NBC and ABC, appeared regularly on *Good Morning America*, and became a celebrated after-dinner speaker, delivering a brilliant account of his remarkable Olympic achievement. This success propelled Jenner to stardom on television and in films.

Initially, Jenner identified himself publicly as a man, but in April 2015 he announced that he was a transwoman and changed his name from Bruce to Caitlyn. She appeared on the July 2015 cover of *Vanity Fair* magazine and starred in a television series called *I Am Cait*, which focused on her gender transition. The name Caitlyn and the gender change became official on September 25, 2015. Jenner described her life in the following terms: "Imagine denying your core and soul. Then add to it the almost impossible expectations that people have for you because you are the personification of The American Male Athlete."[10] After revealing her true self, Caitlyn became an executive producer of *I Am Cait*, which won acclaim for increasing public awareness of transgender issues.

TRANSGENDER CHILDREN AND ADOLESCENTS

For transgender children who think their bodies are the wrong sex, puberty can be profoundly confusing and distressing, as it was for Ben Barres. To alleviate this psychological trauma, physicians are increasingly giving transgender adolescents drugs to block puberty until their bodies and their decision-making capabilities are mature enough to begin cross-sex hormone treatment, typically at age sixteen. But the side effects of these drugs are still largely unknown.

A U.S. study now under way may offer some clarification of when and how best to help adolescents who are seeking to transition from the sex they were assigned at birth. The study, funded by the National Institutes of Health, aims to recruit about three hundred adolescents who have identified themselves as transgender and to follow them for at least five years. The project will be the largest study of transgender youth thus far, and only the second study to track the psychological effects of delaying puberty. It will also be the first study to track the medical impacts of delaying puberty. One group will receive puberty blockers at the beginning of adolescence; another, older group will receive cross-sex hormones.

By the time they reach puberty, 75 percent of children who have questioned their gender will identify as the gender assigned at birth. However, those who identify as transgender in adolescence almost always do so permanently. Some people question the idea of providing puberty-blocking drugs when their side effects are not well understood. Yet denying transgender adolescents the ability to transition by withholding the drugs is unethical, say many people who are involved in this area of treatment. Failing to treat adolescents is not simply being neutral, they point out; it means exposing them to harm.

The Endocrine Society is working to update its guidelines for treating transgender youth. Stephen Rosenthal, a pediatric endocrinologist at the University of California, San Francisco, and a leader of the effort, expects that the guidelines, which now advise clinicians to withhold cross-sex hormone therapy until age sixteen, will allow greater flexibility, since many children enter puberty before age sixteen. Another change in guidelines may encourage children to live as the gender they identify with before puberty. This is an increasingly popular choice, says Diane Ehrensaft, a psychologist at the University of California, San Francisco, but it is controversial.[11] Many psychologists discourage such social transitioning until the teenage years.

No matter what the approach to children's gender identity, says bioethicist Simona Giordano of the University of Manchester, clinicians and families should help children to understand what they are experiencing. "Going through the social and physical transition is a long journey."[12]

LOOKING AHEAD

Sexual differentiation of the brain is a rich and important field of study that is beginning to uncover the neural circuits governing gender-specific behavior, including cognitive aspects of behavior such as gender identity. We now realize, for example, that gender identity has a biological basis and that it can diverge from anatomical sex during prenatal development. Moreover, as Swaab and Garcia-Falgueras note, "There is no proof that social environment after birth has an effect on gender identity or sexual orientation."[13]

A sharper focus on the biology of gender identity will give us a much clearer picture of the range of human sexuality and thus make us more understanding and accepting of transgender men and women. It will enable us to understand what a child means when he or she declares, "I am in the wrong body." And it will enable us to help that child transition into adulthood.

11

CONSCIOUSNESS: THE GREAT REMAINING MYSTERY OF THE BRAIN

Francis Crick, the most important biologist of our time, devoted the last thirty years of his life to studying how consciousness arises from the workings of the brain. "[Y]our joys and your sorrows, your memories and your ambitions, your sense of personal identity and free will, are in fact no more than the behavior of a vast assembly of nerve cells and their associated molecules," Crick wrote in his 1994 book *The Astonishing Hypothesis: The Scientific Search for the Soul.*

Crick made relatively little progress in figuring out the mechanisms of consciousness, however, and today the unity of consciousness—our awareness of self—remains the greatest mystery of the brain. As a philosophical concept, consciousness continues to defy consensus, but most people who study it, and who have examined disorders of consciousness, think of it not as a unitary function of mind but as different states in different contexts.

One of the most surprising insights to emerge from the modern study of states of consciousness is that Sigmund Freud was right: we cannot understand consciousness without understanding that complex, unconscious mental processes pervade conscious thought. All conscious perception depends on unconscious processes. So as we delve into the mystery of consciousness, let us remember what our exploration of brain disorders has taught us about mental processing. We know that the brain uses unconscious and conscious processes to construct an internal representation of the outside world that guides our behavior and our thoughts. If the neural circuits of our brain are disordered, we experience the world differently in both degree and kind than other people do, on both conscious and unconscious levels.

The new biology of mind—the marriage of modern cognitive psychology and neuroscience—has created a new understanding of consciousness. As we shall see in this chapter, scientists have used brain imaging to explore different states of consciousness, revealing some basic ways in which our brain gives rise to our mind. Next, we revisit decision making, this time not from the perspective of faulty moral decision making, but from the broader perspective of how this critical skill makes use of both unconscious and conscious processing. Along the way, we learn what the unlikely collaboration of economics and cell biology has revealed about the rules that govern decision making. Finally, we consider the contributions of psychoanalysis to our understanding of mental processes, and how this mode of treatment can derive renewed power and purpose by engaging with the new biology of mind.

FREUD'S VIEW OF THE MIND

Freud divided our mind into conscious and unconscious components. The conscious mind, the *ego*, is in direct contact with the outside world through our sensory systems for vision, hearing, touch, taste, and smell. The ego is guided by reality, what Freud called the *reality principle*, and is concerned with perception, reasoning, the planning of actions, and the experiencing of pleasure and pain, qualities that enable us to defer gratification. Freud later realized that the ego also has an unconscious component, as we shall see.

The unconscious mind, the *id*, is not governed by logic or reality but by the *pleasure principle*—that is, by seeking pleasure and avoiding pain. Freud initially defined the unconscious as a single entity consisting largely of instincts that lie outside our awareness yet influence our behavior and our experience. He considered instincts to be the principal motivating forces in all mental functions. While Freud held that an infinite number of such instincts exist, he reduced them to a basic few, which he divided into two broad groups. *Eros*, the life instinct, covers all self-preservation and erotic instincts; *Thanatos*, the death instinct, covers all aggressive, self-destructive, and cruel instincts. Thus it is incorrect to think of Freud as asserting that all human actions spring from sexual motivation. Those that spring from Thanatos are not sexually motivated; moreover, as we shall see, the life and death instincts can be fused.

Freud later expanded his idea of the unconscious mind beyond the id, or instinctual unconscious. He added a second component, the *superego*. The superego is the ethical component of mind that forms our conscience. Freud completed his structural model of the mind by adding a third component, the *preconscious unconscious*, which is now called the *adaptive unconscious*. This third unconscious component is part of the ego; it processes the information necessary for consciousness without our being aware of it (fig. 11.1). Thus, Freud appreciated that a great deal of our higher cognitive processing occurs unconsciously, without awareness and without the capacity to reflect. We will return to the adaptive unconscious and its role in decision making later in this chapter.

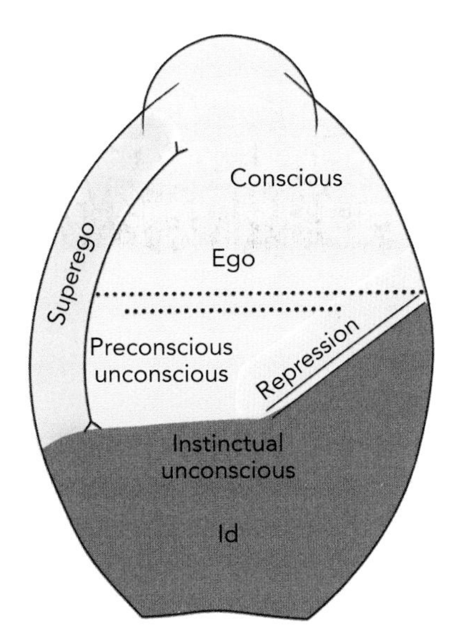

Figure 11.1. Freud's structural model of the mind

Much of Freud's work was devoted to the id, our unconscious storehouse of socially unacceptable desires, traumatic memories, and painful emotions, and to the study of repression, the defensive mechanism that keeps these emotions from entering our conscious thought. Brain scientists are now beginning to examine the biological basis of some of our instincts, the powerful subterranean forces that shape our motivations, behavior, and decision making.

In studies of the neurobiology of emotional behavior, David Anderson of the California Institute of Technology, whom we first encountered in chapter 10, has found some of the biological underpinnings of two of the instincts that Freud observed—eroticism and aggression—as well as the fusion of those instincts.[1]

We have known for some time that the amygdala orchestrates emotion and that it communicates with the hypothalamus, the region that controls instinctive behavior such as parenting, feeding, mating, fear, and fighting (fig. 11.2). Anderson has found a nucleus, or cluster of neu-

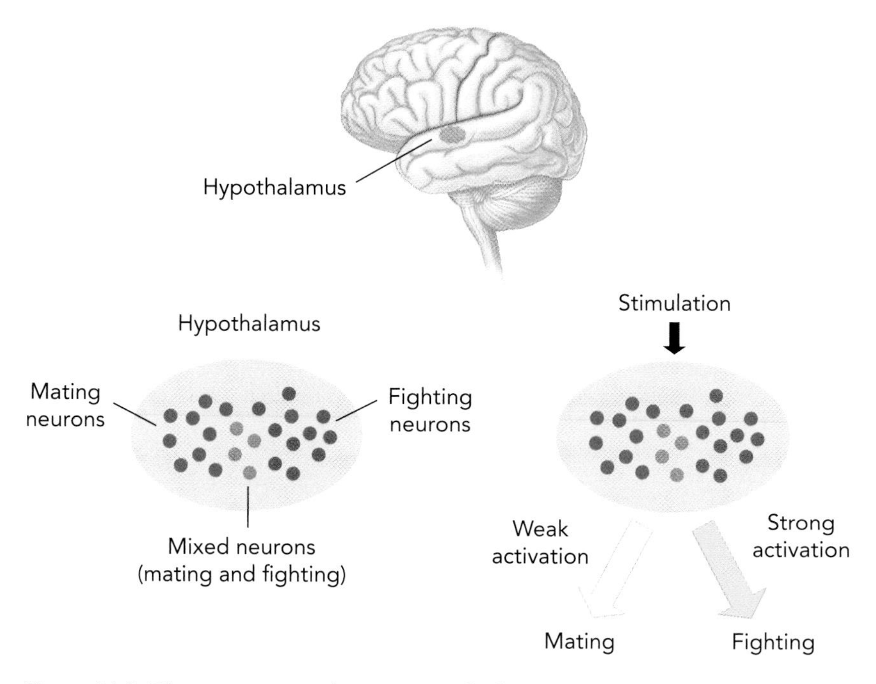

Figure 11.2. The two groups of neurons in the hypothalamus that regulate mating and fighting are closely linked.

rons, within the hypothalamus that contains two distinct populations of neurons: one that regulates aggression and one that regulates sex and mating. About 20 percent of the neurons located on the border between the two populations can be active during either mating or aggression. This suggests that the brain circuits regulating these two behaviors are intimately linked.

How can two mutually exclusive behaviors—mating and fighting—be mediated by the same population of neurons? Anderson found that the difference hinges on the intensity of the stimulus applied to those neurons. Weak sensory stimulation, such as foreplay, activates mating, whereas stronger stimulation, such as danger, activates fighting.

The proximity of the regions concerned with sexuality and aggression, and the zone of overlap, help explain why these two instinctual drives can be so readily fused, as they are, for example, in sexual rage, the extra pleasure some couples derive from sexual experiences that follow an argument.

THE COGNITIVE PSYCHOLOGICAL VIEW OF CONSCIOUSNESS

Modern cognitive psychology has taken an approach to mind that differs from Freud's. Rather than focusing on our instincts, it has focused on how our unconscious mind makes possible a variety of cognitive processes without our being aware of them. But before examining unconscious cognition, let us first consider how modern cognitive psychologists view consciousness.

When cognitive psychologists refer to consciousness, they are talking about different states in different contexts: awakening from sleep, being aware of an oncoming person, sensory perception, and the planning and execution of voluntary action. To understand these different states, we must analyze our conscious experience from two independent but overlapping perspectives.

The first perspective is the *overall arousal state of the brain*—for example, being awake versus being in deep sleep. From this perspective, *level of consciousness* refers to different states of arousal and vigilance, from awakening from sleep to alertness to normal conscious thought, whereas *lack of consciousness* refers to conditions such as sleep, coma, and general anesthesia.

The second perspective is the *content of processing in the aroused state of the brain*—for example, feeling hungry, seeing a dog, or smelling cinnamon. From the perspective of content, we need to determine what aspects of sensory information are processed consciously and what aspects are processed unconsciously, as well as the advantages of each type of processing.

These two perspectives are obviously related: unless we are in an appropriate state of wakefulness, we cannot process sensory stimuli, consciously or unconsciously. So we begin by considering the biology of wakefulness.

Until recently, wakefulness—arousal and vigilance—was considered a result of sensory input to the cerebral cortex: when sensory input is turned off, we fall asleep. In 1918 Constantin von Economo, an Austrian psychiatrist and neurologist who was studying the flu pandemic, had several patients who were in a coma before they died. When he carried out autopsies on their bodies, he found that their sensory systems were largely intact, but a region of the upper brain stem was damaged. He called this region the wakefulness center.

Von Economo's finding was tested empirically in 1949 by Giuseppe Moruzzi, an Italian scientist of great renown, and Horace Magoun, a major American physiologist. In experiments on animals, they found that severing the neural circuits that run from the sensory systems to the brain—specifically, the circuits that mediate touch and sense of position—in no way interferes with consciousness, the wakeful state. However, damaging a region of the upper brain stem—von Economo's wakefulness center—produces coma. Moreover, stimulating that region would awaken an animal from sleep.

Moruzzi and Magoun realized that the brain contains a system—which they called the *reticular activating system*—that extends from the brain stem and midbrain to the thalamus, and from the thalamus to the cortex. This system carries the sensory information from the various sensory systems necessary for the conscious state, and distributes it diffusely to the cerebral cortex (fig. 11.3). But while the reticular activating system is necessary for wakefulness, it is not concerned with the content of conscious processing, that is, with the content of awareness.

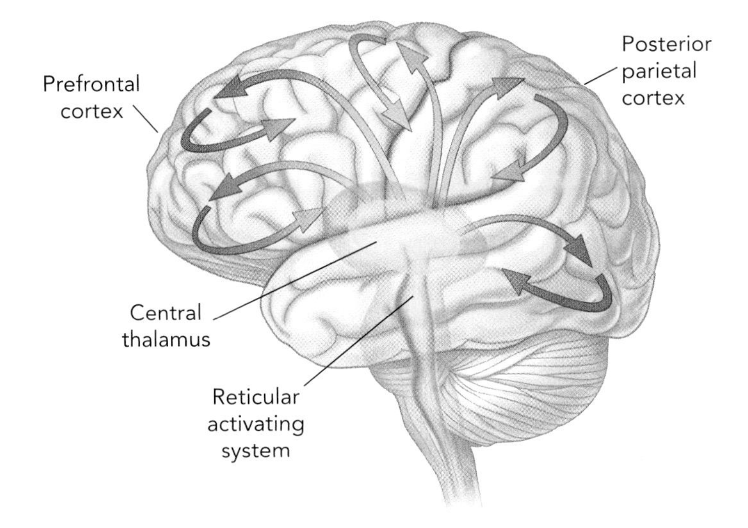

Figure 11.3. The reticular activating system distributes sensory information necessary for the conscious state from the brain stem to the cerebral cortex.

The content of awareness, our conscious state, is mediated by the cerebral cortex. John Searle, a professor emeritus of philosophy at the University of California, Berkeley, argues that although people sometimes say consciousness is hard to define, the commonsense definition is not so difficult. Consciousness is the state of awareness, or sentience. It begins in the morning when we wake up, and it continues all day until we go to sleep again at night, or otherwise become unconscious.

Consciousness has three remarkable features. The first is *qualitative feeling*: listening to music is different from smelling a lemon. The second is *subjectivity*: awareness is going on in me. I am pretty sure that something similar is going on in you, but my relation to my own consciousness is not like my relation to anybody else's. I know you are feeling pain when you burn your hand, but that's because I am observing your behavior, not because I am experiencing—actually feeling—your pain. Only when I burn myself do I feel pain. The third feature is *unity of experience*: I experience the feeling of my shirt against my neck and the sound of my voice and the sight of all the other people sitting around the table as part of a single, unified consciousness—*my* experience—not a jumble of discrete sensory stimuli.

Searle goes on to say that there is an easy problem of consciousness and a hard problem. The easy problem is figuring out what biological processes in the brain correlate with our conscious state. At present, scientists such as Bernard Baars and Stanislas Dehaene are beginning to look for such *neural correlates of consciousness* using brain imaging and a variety of other modern techniques. We shall return to their work later.

The hard problem, according to Searle, is figuring out how these neural correlates of our conscious state relate to conscious experience. We know that every experience we have—the smell of a rose, the sound of a Beethoven piano sonata, the angst of postindustrial man under late capitalism, everything—is produced by variable rates of firing of the neurons in our brain. But do these neural processes, these correlates of consciousness, actually *cause* consciousness? If so, how? And why does conscious experience require these biological processes?

In theory, we should be able to determine whether neural correlates cause consciousness by the usual methods: see if consciousness can be turned on by turning on the neural correlates of consciousness, and see if consciousness can be turned off by turning off the neural correlates of consciousness. We're not quite able to do that yet.

THE BIOLOGY OF CONSCIOUSNESS

The nineteenth-century physiologist and psychologist Hermann von Helmholtz was probably the first person to realize that the brain assembles basic bits of information from our sensory systems and unconsciously draws inferences from them. In fact, the brain can make complex inferences from very scanty information. When you look at a series of black lines, for instance, the lines don't mean anything; but if the lines begin to move—and particularly if they move forward—your brain instantly recognizes them as a person walking.

Helmholtz also realized that unconscious processing of information isn't just reflexive or instinctive, it's adaptive—it helps us to survive in the world. Moreover, our unconscious is creative. It integrates a range of information and delivers it to consciousness, using both information that is stored in memory and information that is currently being perceived. The brain takes this partial information, compares it to previous experience, and makes a more learned, rational judgment.

This was an amazing insight, and Freud picked up on it. He was interested in a group of diseases called the *aphasias*, various defects in the capacity to speak, and he made a remarkable observation: we don't consciously pick the words we're going to use. We don't consciously form grammatical structure. It's all done unconsciously—we just speak. In fact, when we speak, we know the gist of what we're going to say, even though we don't know precisely what we're going to say until we say it.

Similarly, when we look at a face, we don't consciously see two eyes and two eyebrows and two ears and a mouth and say, "Ah, yes, that's so-and-so." Recognition just comes to us. Such high-level, adaptive thinking takes place in Freud's preconscious unconscious. Thus, Freud's question might really be, "What is the nature of all that integration that allows us to recognize something complex?"

To answer this question, look at figure 11.4. The left side seems to show a white square lying on top of four black disks; the right side seems

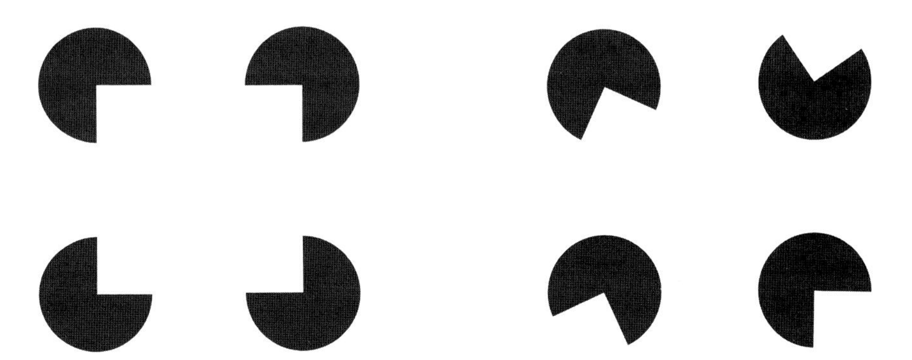

Figure 11.4. The Kanizsa Square: conscious thought creates implied lines (*left*) where actual lines do not exist.

to show four black disks, each with a piece taken out of it. Your brain, which is accustomed to making sense out of its perceptual experience, tells you that you are looking on the left at a white square lying on top of four black disks. But in fact that white square isn't there. Your brain has created it. When you look at the four black disks on the right, you realize this. Moreover, your brain even creates a difference between the whiteness of the square lying on the disks and the whiteness of the background, a difference that also doesn't exist.

The cognitive psychologist Bernard Baars thought that the brain's integration of conscious and unconscious mental processes—our mind's interpretation of what we see—could probably be explored empirically if it could somehow be linked to developments in neuroscience. He therefore set out to do that.

THE GLOBAL WORKSPACE

After designing and conducting a series of experiments that used brain imaging to study visual perception, Baars introduced the theory of *the global workspace* in 1988.[2] According to this theory, consciousness involves the widespread dissemination, or broadcasting, of previously unconscious (preconscious) information throughout the cortex. Baars suggested that the global workspace comprises a system of neural circuits that extends from the brain stem to the thalamus and from there to the cerebral cortex.

Before Baars, the question of consciousness was taboo among most rigorous experimental psychologists because it was not considered a problem that could be examined scientifically. We now realize, however, that psychology has an enormous variety of techniques at its disposal for examining consciousness in the laboratory. Basically, an experiment can take any stimulus—an image of a face or a word—change the conditions a bit, and make our perception of it come into and go out of consciousness at will. For example, if I show you a photograph of a person's face followed very quickly by a different image, which *masks* the face, you will not consciously perceive the face. But if I show you the same photograph for several seconds, you will readily perceive it consciously.

This was a new, cognitive psychological understanding of consciousness. It synthesized the psychology of conscious perception and the brain science of neural signals being broadcast from the thalamus to the entire cerebral cortex. The two approaches are inseparable. Without a good psychology of the conscious state we can't make progress in the biology of broadcasting information, and without the biology we will never understand the underlying mechanism of consciousness.

The French cognitive neuroscientist Stanislas Dehaene extended Baars's psychological model to a biological model.[3] Dehaene found that what we experience as a conscious state is the result of a distributed set of neural circuits that select a piece of information, amplify it, and broadcast it forward to the cortex. Baars's theory and Dehaene's findings show that we have two different ways of thinking about things: one is unconscious, involving perception; the other is conscious, involving the broadcasting of the perceived information.

Dehaene devised a way to image consciousness in the brain by contrasting unconscious and conscious processing.[4] He flashes the words "one, two, three, four" on a screen. Even when he flashes them very quickly, you can see them. But when he flashes a shape just before and just after the last word, "four," the word seems to disappear. It is still there on the screen, it's still there on your retina, your brain is processing it—but you are not conscious of it.

Going a little bit further, he then places the words just at the threshold of consciousness, so that half of the time you will say you saw them, and half of the time you will say you didn't see them. Your perception is purely subjective. The objective reality of the words is exactly the same, whether you think you saw them or not.

What happens in the brain when we see a word subliminally, below the threshold of consciousness? First, the visual cortex becomes very active. This is unconscious neural activity: the word we've seen reaches the early visual processing station of the cerebral cortex. Then after 200 or 300 milliseconds it slowly dies out without reaching the higher centers of the cortex (fig. 11.5). Thirty years ago, if asked whether an unconscious perception reaches the cerebral cortex, neuroscientists would have said no, because they believed that any information reaching the cerebral cortex would automatically enter consciousness. In fact, however, when a perception becomes conscious, something quite different occurs.

Conscious perception also begins with activity in the visual cortex, but instead of dying out, the activity is amplified. After about 300 milliseconds, it becomes very large: it's like a tsunami instead of a dying wave. It propagates higher into the brain, up to the prefrontal cortex.

From there it goes back to where it started, creating a reverberating neural circuit of activity. This is the broadcasting of information that occurs when we are conscious of that information. It moves information into the global workspace, where it is accessible to other regions of the brain (fig. 11.6).

Put simply, when you are conscious of a particular word, that word becomes available in the global workspace, a process that takes place separately from your visual recognition of the word. Although the word is flashed in front of your eyes for only a very brief moment, you can keep that word in mind with your working memory. You can then broadcast it to all of the areas that need it.

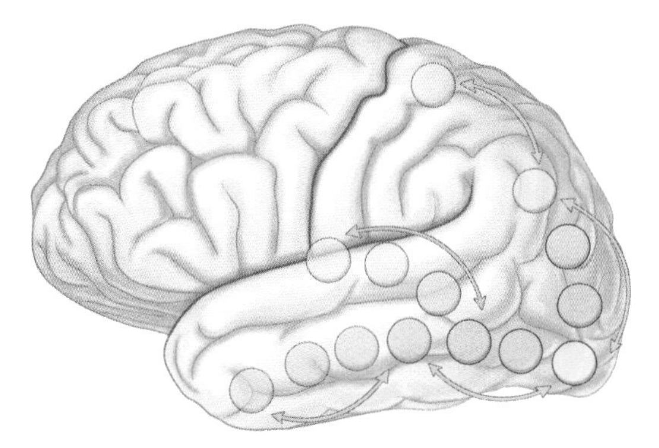

Figure 11.5. Subliminal perception: activity in the visual cortex dies out before reaching higher regions of the brain.

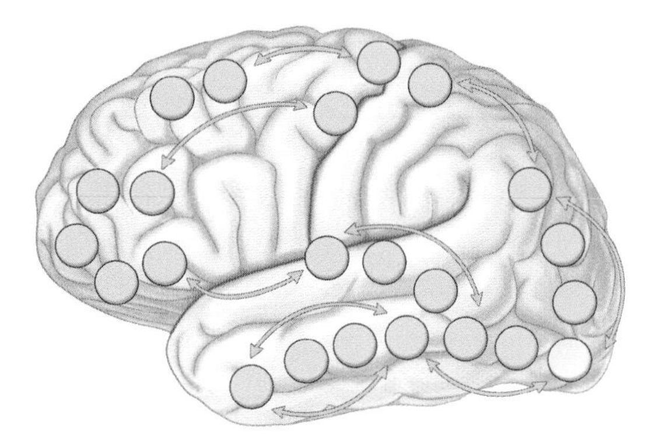

Figure 11.6. Conscious perception: activity in the visual cortex is broadcast to the prefrontal cortex, where it is available to other regions of the brain.

The basic finding from brain imaging is the same. Conscious activity is restricted in what it can focus on: it selects only a single item at a time and broadcasts it widely across the brain. Unconscious processing of information, in contrast, can take place in many different areas of the cortex simultaneously, but that information is not broadcast to other areas. As you read these words, for example, you are aware of your surroundings—ambient sounds, temperature, and so on. That sensory information about your surroundings is processed unconsciously in the brain, but because the information is not broadcast widely, you are not consciously aware of it as you are reading.

The experiments previously described demonstrate that information can enter our brain yet not give rise to conscious perception. Intriguingly, however, such information *can* affect our behavior, as we shall see. That is because the brain's unconscious processing isn't limited to sensory information. While the mere recognition of a word is occurring unconsciously, the meaning of that word is being accessed at much higher levels in the brain without our being aware of it. Other aspects of the word can also be computed unconsciously, such as its sound, or its emotional content, or whether we spoke it in error and want to catch the error. Similarly, when we see a number, we effortlessly tap into the mathematical systems of our brain. Scientists are still struggling to understand how unconscious processing works and how deep it can go.

CORRELATION OR CAUSATION?

How do we distinguish between something that is preconscious and thus correlated with conscious activity (the neural correlate of consciousness) and something that actually *causes* conscious activity? How does the brain encode the actual content of consciousness? To make progress on these questions we will need more fine-grained techniques.

Daniel Salzman, now at Columbia University, and William Newsome of Stanford University have used electrical stimulation to manipulate the information-processing pathways in the brains of animals.[5] The animals are trained to indicate whether dots on a screen are moving to the left or to the right. By stimulating just a tiny bit of the brain area that is concerned with visual movement, Salzman and Newsome can induce a slight change in the animals' perception of which way the dots are moving. This change in perception causes the animals to change their minds about which way the dots are moving. Thus, the dots might actually be moving right, but when Newsome and Salzman stimulate brain cells that care about leftward movement, the animals change their minds and indicate that the dots are moving left.

In parallel work in 1989, Nikos Logothetis and Jeffrey Schall examined *binocular rivalry*.[6] Binocular rivalry describes the situation in which one image is presented to one eye and a different image is presented to the other eye. Instead of the two images being superimposed, our perception flips from one image to the other: we are only aware of "seeing" one image at a time. The same phenomenon occurs in animals. In their experiments, Logothetis and Schall trained monkeys to "report" these flips. They found that some neurons respond only to the physical image, while others respond to the animal's perception of it. As we have seen, perception involves cognitive functions, such as memory, not simply responses to sensory stimuli. Logothetis and Schall's study has spawned additional work, the gist of which is that the number of neurons attuned to *percepts*, or mental representations of an object, becomes greater as information moves from the primary visual cortex to higher regions of the brain.

Logothetis concludes from his and related work: "The picture of the brain that starts to emerge from these studies is of a system whose processes create states of consciousness in response not only to sensory inputs but also to internal signals representing expectations based on past experiences."[7] He goes on to state that "our success in identifying neurons that reflect consciousness is a good start" toward uncovering the neural circuits underlying consciousness.

Although we are only beginning to study the biology of consciousness, these experiments have given us some useful paradigms for exploring different states of consciousness.

AN OVERALL PERSPECTIVE ON THE BIOLOGY OF CONSCIOUSNESS

It is tempting to conclude that the propagation of electrical signals forward into the prefrontal cortex—the broadcasting of unconscious information to the global workspace—represents consciousness, but consciousness is not likely to be that simple. Some of this broadcast activity does represent consciousness, but some of it may just represent associations.

Suppose, for example, that someone who does not know who John Lennon was looked at a photograph of him. That person's brain would go through the usual process of sending information from the visual cortex to the prefrontal cortex; as a result, she would see a pleasant-looking guy with round glasses and long hair. However, if that person did know who John Lennon was, she might associate Lennon's image with the song "Eleanor Rigby" and with Paul McCartney, George Harrison, and Ringo Starr, the other Beatles. That additional brain activity is separate from perception of Lennon's face: it recognizes the image of Lennon and associates it with memories. We make those associations unconsciously, but they nevertheless result from activity in the frontal areas of our brain in response to information sent from the visual system.

One final, very important point concerns the fact that consciousness can operate largely independently of incoming stimuli. We generally envision the brain as receiving sensory input and producing outputs in response. That is often accurate, but consider this: even in complete darkness, with no visual stimuli, we maintain very complex states of activity that originate high up in the cortex and thus are top-down, or cognitive, in nature. Moreover, when we dream, we may be aware of highly colorful and emotionally arousing events, even though some signals from the external world may be blocked from reaching the cortex. Sometimes we think and plan while ignoring external events around us. Even when we daydream, imagining future events, our brain temporarily blocks sensory stimuli and instead plays with our internally generated ideas. These ideas and daydreams are generated independently, without input from external stimuli. To be sure, our brain can be brought down to earth by a loud noise or the smell of smoke, but while we are concentrating on our inner thoughts—as we often do—our brain keeps new sensory stimuli out.

DECISION MAKING

The ability to make good decisions is a critical skill that depends on both unconscious and conscious mental processing. In chapter 8 we discussed the important role of emotion in decision making. Here, we go beyond that to explore several ideas from cognitive psychology and biology that have advanced our understanding of how conscious and unconscious processes interact in decision making.

Timothy Wilson, a cognitive psychologist, introduced the idea of the *adaptive unconscious*, a set of high-level cognitive processes similar to Freud's preconscious unconscious.[8] The adaptive unconscious interprets information quickly, without our being aware of it, which makes it vital to our survival. While we consciously focus on what's happening around us, the adaptive unconscious lets part of our mind keep track of what's going on elsewhere, to make sure we don't miss something important. The adaptive unconscious serves a number of functions, one of which is decision making.[9]

Many of us, when faced with an important choice, take out the proverbial piece of paper and make a list of plusses and minuses to help us decide what to do. But experiments have shown that this may not be the best way to make a decision. If you are overly conscious, you may talk yourself into thinking you prefer something that you really don't like. Instead, you are best off when you allow yourself to gather as much information as possible about the decision and then let it percolate unconsciously. A preference will bubble up. Sleeping helps equilibrate emotions, so when it comes to an important decision, you should literally sleep on it. So there it is: our conscious decisions rest on information that is selected from our unconscious.

Although the adaptive unconscious is a very smart, sophisticated set of processes, it isn't perfect. It categorizes very quickly and can be a little rigid. One school of thought holds that this may account, in part, for prejudice. We react to a stimulus quickly, on the basis of past experience that may not apply to the new situation at hand. In such new situations, consciousness may step in and correct a snap judgment, saying, "Wait a minute. This quick, negative reaction may be wrong. I need to rethink it." The adaptive unconscious works in tandem with consciousness to guide us in ways that make us the smartest species on earth. It would be interesting to see how far back we could trace these two mental processes that evolved to deal with different kinds of information.

The biological role of the adaptive unconscious in decision making was revealed in a simple experiment by Benjamin Libet at the University of California, San Francisco. Hans Helmut Kornhuber, a German neurologist, had shown that when we initiate a voluntary movement, such as moving a hand, we produce a *readiness potential*, an electrical signal that can be detected on the surface of the skull. The readiness potential appears a split second before the actual movement.

Libet carried this experiment a step further. He asked people to consciously "will" a movement and to note exactly when that willing occurred. He was sure it would occur before the readiness potential, the signal that activity had begun. What he found, to his surprise, was that it occurred *after* the readiness potential. In fact, by averaging a number of trials, Libet could look into a person's brain and tell that he or she was about to move even before the person was aware of it.[10]

This astonishing result might suggest that we are at the mercy of our unconscious instincts and desires. In fact, however, the activity in our brain precedes the *decision* to move, not the movement itself. As Libet explains, the process of initiating a voluntary action occurs rapidly in an unconscious part of the brain; however, just before the action is begun, consciousness, which comes into play more slowly, approves or vetoes the action. Thus, in the 150 milliseconds before you lift your finger, your consciousness determines whether or not you will actually move it. What Libet showed is that activity in the brain precedes awareness, just as it precedes any action we take. We therefore have to refine our thinking about the nature of brain activity as it pertains to consciousness.

In the 1970s Daniel Kahneman and Amos Tversky began to entertain the idea that intuitive thinking functions as an intermediate step between perception and reasoning. They explored how people make decisions and, in time, realized that unconscious errors of reasoning greatly distort our judgment and influence our behavior.[11] Their work became part of the framework for the new field of behavioral economics.

Kahneman and Tversky identified certain mental shortcuts that, while allowing for speedy action, can result in less-than-optimal judgments. For example, decision making is influenced by the way choices are described, or *framed*. In framing a choice, we weigh losses far more heavily than equivalent gains. If a patient needs surgery, for instance, he is far more likely to undergo the procedure if the surgeon says that 90 percent of patients survive perfectly well, as opposed to saying that 10 percent of patients die. The numbers are the same, but because we are averse to risk, we much prefer to hear that we have a high probability of living than that we have a low probability of dying.

Kahneman went on to describe two general systems of thought.[12] *System 1* is largely unconscious, fast, automatic, and intuitive—like the adaptive unconscious, or what Walter Mischel, a leading cognitive psychologist, calls "hot" thinking. In general, System 1 uses association and metaphor to produce a quick rough draft of an answer to a problem or situation. Kahneman argues that some of our most highly skilled activities require large doses of intuition: playing chess at a master level or appreciating social situations. But intuition is prone to biases and errors.

System 2, in contrast, is consciousness-based, slow, deliberate, and analytical, like Mischel's "cool" thinking. System 2 evaluates a situation using explicit beliefs and a reasoned evaluation of alternatives. Kahneman argues that we identify with System 2, the conscious, reasoning self that makes choices and decides what to think about and what to do, whereas actually our lives are guided by System 1.

A clear example of the systems biology of decision making has emerged from the study of unconscious emotion, conscious feeling, and their bodily expression. Until the end of the nineteenth century, emotion was thought to result from a particular sequence of events: a person recognizes a frightening situation; that recognition produces a conscious experience of fear in the cerebral cortex; and the fear induces unconscious changes in the body's autonomic nervous system, leading to increased heart rate, constricted blood vessels, increased blood pressure, and moist palms.

In 1884, as we have seen, William James set this sequence of events on its ear. James realized not only that the brain communicates with the body but, equally important, that the body communicates with the brain. He proposed that our conscious experience of emotion takes place *after* the body's physiological response. Thus, when we encounter a bear sitting in the middle of our path, we do not consciously evaluate the bear's ferocity and then feel afraid; we instinctively run away from it and only later experience conscious fear.

Recently, three independent research groups have confirmed James's theory.[13] Using brain imaging, they discovered the anterior insular cortex, or insula, a little island in the cortex located between the parietal and temporal lobes. The insula is where our feelings are represented—our conscious awareness of the body's response to emotionally charged stimuli. The insula not only evaluates and integrates the emotional or motivational importance of these stimuli, it also coordinates external sensory information and our internal motivational states. This consciousness of bodily states is a measure of our emotional awareness of self, the feeling that "I am."

Joseph LeDoux, a pioneer in the neurobiology of emotion whom we met in chapter 8, found that a stimulus takes one of two routes to the amygdala. The first is a rapid, direct pathway that processes unconscious sensory data and automatically links the sensory aspects of an event together. The second pathway sends information through several relays in the cerebral cortex, including the insula, and may contribute to the conscious processing of information. LeDoux argues that together, the direct and indirect pathways mediate both the immediate, unconscious response to a situation and the later, conscious elaboration of it.

With these studies, we are now in a position to go beneath the surface of mental life and begin to examine how conscious and unconscious experiences are related. In fact, some of the most fascinating recent insights into consciousness have come from studies that parallel James's thinking and examine consciousness through its role in other mental processes. Imaging studies by Elliott Wimmer and Daphna Shohamy, for instance, show that the same mechanisms in the hippocampus that are involved in the conscious recall of a memory can also guide and bias unconscious decisions.[14]

Wimmer and Shohamy designed a study in which they first showed participants a series of paired images. The scientists then separated the images and, using the techniques of conditioned learning, presented some of the images to participants together with a monetary reward. Finally, they showed participants the images that had *not* been linked to a monetary reward and asked them which of the latter images they preferred. Participants tended to prefer images that had previously been paired with a rewarded image, even though the participants could not consciously recall the original pairs. The researchers concluded that the hippocampus can reactivate the association of the current image with its original mate and, working with the striatum, connect it to the memory of the reward, thus biasing a participant's choice.

Following on the realization that biology is involved in decision making and choice, Newsome and other neuroscientists began applying such economic models on the cellular level in animals in an effort to understand the rules that govern decision making. Meanwhile, economists began to incorporate the outcomes of those studies into their theories of economics.

Neuroscientists have made good progress in studies of decision making by examining single nerve cells in primates. A key finding, epitomized by the work of Michael Shadlen, is that neurons in the association areas of the cortex, which are involved in decision making, have very different response properties than neurons in the sensory areas of the cortex. Sensory neurons respond to a current stimulus, whereas association neurons are active longer, presumably because they are part of the mechanism that links perception with a provisional plan for action.[15]

Shadlen's results indicate that association neurons accurately track the probabilities related to making a choice. For example, as a monkey sees more and more evidence indicating that a rightward target will dispense a reward, the neural activity that favors a rightward choice increases. This allows the monkey to accumulate evidence and make a choice when the probability of being correct passes some threshold, say 90 percent. The neurons' activity and the decision they drive can occur very rapidly—often in less than a second. Thus, under the right circumstances, even rapid decisions can be made in nearly optimal fashion. This may explain why the fast, unconscious, System 1 mode of thinking has survived: it may be prone to error under some circumstances, but it is highly adaptive under others.

PSYCHOANALYSIS AND THE NEW BIOLOGY OF MIND

During the first half of the twentieth century, psychoanalysis provided remarkable new insights into unconscious mental processes, psychic determinism, infantile sexuality, and, perhaps most important of all, the irrationality of human motivation. Its approach was so novel and so powerful that for many years not only Freud but other intelligent and creative psychoanalysts as well could argue that psychotherapeutic encounters between patient and analyst provided the best context for scientific inquiry into the human mind.

But the achievements of psychoanalysis during the second half of the century were less impressive. Although psychoanalytic thinking continued to progress, there were relatively few brilliant new insights. Most important, and most disappointing, psychoanalysis did not evolve scientifically. Specifically, it did not develop objective methods for testing the exciting ideas it had formulated. As a result, psychoanalysis entered the twenty-first century with its influence in decline.

What led to this regrettable decline? First, psychoanalysis had exhausted much of its investigative power. Freud listened carefully to patients, and he listened in new ways. He also presented a provisional schema for making sense out of their apparently unrelated and incoherent associations. Today, however, little that is new in the way of theory remains to be learned merely by listening carefully to individual patients. Moreover, clinical observation of individual patients, in a context as susceptible to observer bias as the psychoanalytic relationship, is not a sufficient basis for a science of mind.

Second, although psychoanalysis often thought of itself as a scientific discipline, it has rarely used scientific methods, and it has failed over the years to submit its assumptions to testable experimentation. Indeed, psychoanalysis has traditionally been far better at generating ideas than at testing them. In part that is because, with rare exceptions, the data gathered in psychoanalytic sessions are private: the patient's comments, associations, silences, postures, movements, and other behaviors are privileged. In fact, privacy is central to establishing the trust needed in a psychoanalytic situation. As a result, we usually have only the analysts' subjective accounts of what they believe happened in sessions. Such accounts are not comparable to scientific data.

Third, with some notable exceptions, psychoanalysts have not embraced the last fifty years' worth of knowledge about the biology of the brain and its control of behavior.

If psychoanalysis is to regain its intellectual power and influence, as it should, it will need to engage constructively with the new biology of mind. Conceptually, the new biology could provide psychoanalysis with a scientific foundation for future growth. Experimentally, biological insights could serve as a stimulus for research, for testing specific ideas about how brain processes mediate mental processes and behavior. Imaging studies have provided evidence that psychoanalysis, as well as other forms of psychotherapy, is a biological treatment—it actually produces detectable, lasting physical changes in the brain and in behavior. Now we need to find out how.

Fortunately, some people in the psychoanalytic community realized that empirical research was essential to the future of the discipline. Because of them, two trends have gained momentum in the last several decades. The first is the effort mentioned above, to align psychoanalysis with the new biology of mind. The second is the insistence on evidence-based psychotherapy, which we considered in chapter 3. Since almost every mental function requires the interplay of conscious and unconscious processes, the new biology of mind can provide a valuable link between psychoanalytic theory and modern cognitive neuroscience. Such a link would enable cognitive neuroscience to explore, modify, and, where appropriate, disprove psychoanalytic theories about the unconscious. It would also enable psychoanalytic ideas to enrich cognitive neuroscience.

Using Dehaene's operational approach, we might explore, for example, how Freud's instinctual unconscious maps onto modern biological insights into social behavior and aggression. Do these unconscious processes reach the cerebral cortex, even though they may not reach consciousness? What neural systems govern mechanisms of defense, such as sublimation, repression, and distortion?

Twenty-first-century biology is already in a good position to answer some of our questions about the nature of conscious and unconscious mental processes, but those answers will be richer and more meaningful if they are reached through a synthesis of the new biology of mind and psychoanalysis. This synthesis would add greatly to our knowledge of mental disorders and thus to our understanding of the neural circuitry of healthy brain function. New insights into healthy brain function would put us in a better position to understand people with brain disorders and to develop effective treatments for them.

LOOKING AHEAD

Consciousness remains a mystery. We know that it is not static, that states of consciousness vary. Moreover, consciousness entails making unconscious perceptual information available to wide areas of the cerebral cortex, especially the prefrontal cortex, the part of the brain responsible for integrating perception, memory, and cognition. Determining the nature of consciousness—in essence, how we acquire our awareness of self from unconscious activity in the brain—is one of the greatest scientific challenges of the twenty-first century, so answers will not come quickly or easily.

While brain disorders can create disturbances in many aspects of our conscious experience—cognition, memory, mood, social interaction, volition, behavior—most of what we have learned about consciousness from these disorders thus far applies to the interaction of conscious and unconscious processes. That interplay is likely to be critical to our eventual understanding of how consciousness arises.

CONCLUSION: COMING FULL CIRCLE

We have learned more about the brain and its disorders in the past century than we have during all of the previous years of human history combined. Decoding the human genome has shown us how genes dictate the organization of the brain and how changes in genes influence disorders. We have new insights into the molecular pathways that underlie specific brain functions, such as memory, as well as the defective genes that contribute to disorders of those functions, such as Alzheimer's disease. We also know more about the powerful interaction of genes and the environment in causing brain disorders, such as the role of stress in mood disorders and PTSD.

Equally remarkable are recent breakthroughs in brain-scanning technology. Scientists can now track particular mental processes and mental disorders to specific regions and combinations of regions in the brain while a person is alert, with active nerve cells lighting up to create brightly colored maps of brain function. Finally, animal models of disorders have pointed us toward new avenues of research in human patients.

As we have seen, brain disorders result when some part of the brain's circuitry—the network of neurons and the synapses they form—is overactive, inactive, or unable to communicate effectively. The dysfunction may stem from injury, changes in synaptic connections, or faulty wiring of the brain during development. Depending on what regions of the brain they affect, disorders change the way we experience life—our emotion, cognition, memory, social interaction, creativity, freedom of choice, movement, or, most often, a combination of these aspects of our nature.

Thanks in large part to advances in genetics, brain imaging, and animal models, scientists studying brain disorders have confirmed several general principles of how our brain normally functions. For example, imaging studies show that the left and right hemispheres of the brain deal with different aspects of mental functions and that the two hemispheres inhibit each other. Specifically, damage to the left hemisphere can free up the creative capabilities of the right hemisphere. More generally, when one neural circuit in the brain is turned off, another circuit, which was inhibited by the inactivated circuit, may turn on.

Scientists have also uncovered some surprising links between disorders that appear to be unrelated because they are characterized by dramatically different kinds of behavior. Several disorders of movement and of memory, such as Parkinson's disease and Alzheimer's disease, result from misfolded proteins. The symptoms of these disorders vary widely because the particular proteins affected and the functions for which they are responsible differ. Similarly, both autism and schizophrenia involve synaptic pruning, the removal of excess dendrites on neurons. In autism, not enough dendrites are pruned, whereas in schizophrenia too many are. In another example, three different disorders—autism, schizophrenia, and bipolar disorder—share genetic variants. That is, some of the same genes that create a risk for schizophrenia also create a risk for bipolar disorder, and some of the same genes that create a risk for schizophrenia also create a risk for autism spectrum disorders.

The interplay of unconscious and conscious mental processes is critical to how we function in the world. We see this particularly clearly in creativity and decision making. Our innate creativity—in any field—hinges on loosening the bonds of consciousness and gaining access to our unconscious. This is easier for some people than for others. Prinzhorn's schizophrenic artists, with their diminished inhibitions and social constraints, had free access to their unconscious conflicts and desires, whereas the Surrealist artists had to devise ways of tapping into theirs. Decision making is different. Here, we are not aware of our unconscious emotions—or of the need for them. Yet studies have shown that people with damage to regions of the brain involving emotion have great difficulty making decisions.

This new biology of mind has revolutionized our ability to understand the brain and its disorders. But how is the synthesis of modern cognitive psychology and neuroscience likely to affect our lives in the future?

The new biology of mind will lead to radical changes in the way medicine is practiced, in two ways. First, neurology and psychiatry will merge into a common clinical discipline that focuses increasingly on the patient as an individual with particular genetic predispositions to health and disease. This focus will move us toward a biologically inspired, personalized medicine. Second, we will have, for the first time, a meaningful and nuanced biology of the processes in the brain that go awry in brain disorders, as well as the processes that lead to the sexual differentiation of our brain and our gender identity.

It is likely that personalized medicine, with its focus on clinical DNA testing—the search for small genetic differences in individuals—will reveal who is at risk of developing a particular disease and thus enable us to modify the course of that disease through diet, surgery, exercise, or drugs many years before signs and symptoms appear. Currently, for example, newborn babies are screened primarily for treatable genetic diseases, such as phenylketonuria. Perhaps in the not-too-distant future, children at high risk for schizophrenia, depression, or multiple sclerosis will be identified and treated to prevent changes that would otherwise occur later in life. Similarly, middle-aged and older people may benefit from a determination of their individual risk profile for late-onset diseases such as Alzheimer's or Parkinson's. Indeed, DNA testing should also allow us to predict individual responses to drugs, including any side effects they may cause, leading to drugs tailored to the needs of individual patients.

My own work has shown that learning—experience—changes the connections between neurons in the brain. This means that each person's brain is slightly different from the brain of every other person. Even identical twins, with their identical genomes, have slightly different brains because they have been exposed to different experiences. It is very likely that, in the course of illuminating brain function, brain imaging will establish a biological foundation for the individuality of our mental life. If it does, we will have a powerful new way of diagnosing brain disorders and evaluating the outcome of various treatments, including different forms of psychotherapy.

Seen in this light, understanding the biology of brain disorders is part of the continuous attempt of each generation of scholars to understand human thought and human action in new terms. It is an endeavor that moves us toward a new humanism, one that draws on knowledge of our biological individuality to enrich our experience of the world and our understanding of one another.

NOTES

For a general introduction to the biology of the brain, see Eric R. Kandel et al., eds., *Principles of Neural Science*, 5th ed. (New York: McGraw Hill, 2013).

INTRODUCTION

1. René Descartes, *The Philosophical Writing of Descartes*, trans. John Cottingham, Robert Stoothoff, and Dugald Murdoch, vol. 1 (Cambridge, U.K., and New York: Cambridge University Press, 1985).
2. John R. Searle, *The Mystery of Consciousness* (New York: The New York Review of Books, 1997).
3. Charles R. Darwin, *The Expression of the Emotions in Man and Animals* (London: John Murray, 1872).

1. WHAT BRAIN DISORDERS CAN TELL US ABOUT OURSELVES

1. Eric R. Kandel and A. J. Hudspeth, "The Brain and Behavior," in Kandel et al., *Principles of Neural Science*, 5th ed., 5–20.
2. William M. Landau et al., "The Local Circulation of the Living Brain: Values in the Unanesthetized and Anesthetized Cat," *Transactions of the American Neurological Association* 80 (1955): 125–29.
3. Louis Sokoloff, "Relation between Physiological Function and Energy Metabolism in the Central Nervous System," *Journal of Neurochemistry* 29 (1977): 13–26.

2. OUR INTENSELY SOCIAL NATURE: THE AUTISM SPECTRUM

For a general discussion on autism, see Uta Frith et al., "Autism and Other Developmental Disorders Affecting Cognition," in Kandel et al., *Principles of Neural Science*, 1425–40.

1. David Premack and Guy Woodruff, "Does the Chimpanzee Have a Theory of Mind?" *Behavioral and Brain Sciences* 1, no. 4 (1978): 515–26.
2. Simon Baron-Cohen, Alan M. Leslie, and Uta Frith, "Does the Autistic Child Have a 'Theory of Mind'?" *Cognition* 21 (1985): 37–46.
3. Uta Frith, "Looking Back," https://sites.google.com/site/utafrith/looking-back-.
4. Kevin A. Pelphrey and Elizabeth J. Carter, "Brain Mechanisms for Social Perception: Lessons from Autism and Typical Development," *Annals of the New York Academy of Sciences* 1145 (2008): 283–99.
5. Leslie A. Brothers, "The Social Brain: A Project for Integrating Primate Behavior and Neurophysiology in a New Domain," *Concepts in Neuroscience* 1 (2002): 27–51.
6. Stephen J. Gotts et al., "Fractionation of Social Brain Circuits in Autism Spectrum Disorders," *Brain* 135, no. 9 (2012): 2711–25.
7. Cynthia M. Schumann et al., "Longitudinal Magnetic Resonance Imaging Study of Cortical Development through Early Childhood in Autism," *Journal of Neuroscience* 30, no. 12 (2010): 4419–27.
8. Leo Kanner, "Autistic Disturbances of Affective Contact," *The Nervous Child: Journal of Psychopathology, Psychotherapy, Mental Hygiene, and Guidance of the Child* 2 (1943): 217–50.
9. Alison Singer, personal communication, March 24, 2017.
10. Ibid.
11. Erin McKinney, "The Best Way I Can Describe What It's Like to Have Autism," *The Mighty*, April 13, 2015, themighty.com/2015/04/what-its-like-to-have-autism-2/.
12. Ibid.
13. Ibid.
14. Beate Hermelin, *Bright Splinters of the Mind: A Personal Story of Research with Autistic Savants* (London and Philadelphia: Jessica Kingsley Publishers, 2001).
15. Stephan J. Sanders et al., "Multiple Recurrent De Novo CNVs, Including Duplications of the 7q11.23 Williams Syndrome Region, Are Strongly Associated with Autism," *Neuron* 70, no. 5 (2011): 863–85.
16. Thomas R. Insel and Russell D. Fernald, "How the Brain Processes Social Information: Searching for the Social Brain," *Annual Review of Neuroscience* 27 (2004): 697–722.
17. Niklas Krumm et al., "A *De Novo* Convergence of Autism Genetics and Molecular Neuroscience," *Trends in Neuroscience* 37, no. 2 (2014): 95–105.
18. Augustine Kong et al., "Rate of *De Novo* Mutations and the Importance of Father's Age to Disease Risk," *Nature* 488 (2012): 471–75.
19. Guomei Tang et al., "Loss of mTOR-Dependent Macroautophagy Causes Autistic-like Synaptic Pruning Deficits," *Neuron* 83, no. 5 (2014): 1131–43.
20. Mario De Bono and Cornelia I. Bargmann, "Natural Variation in a Neuropeptide Y Receptor Homolog Modifies Social Behavior and Food Response in *C. elegans*," *Cell* 94, no. 5 (1998): 679–89.
21. Thomas R. Insel, "The Challenge of Translation in Social Neuroscience: A Review of Oxytocin, Vasopressin, and Affiliative Behavior," *Neuron* 65, no. 6 (2010): 768–79.
22. Ibid.

23. Sarina M. Rodrigues et al., "Oxytocin Receptor Genetic Variation Relates to Empathy and Stress Reactivity in Humans," *PNAS* 106, no. 50 (2009): 21437–41.
24. Simon L. Evans et al., "Intranasal Oxytocin Effects on Social Cognition: A Critique," *Brain Research* 1580 (2014): 69–77.
25. Tang et al., "Loss of mTOR-Dependent Macroautophagy."

3. EMOTIONS AND THE INTEGRITY OF THE SELF: DEPRESSION AND BIPOLAR DISORDER

1. William Styron, *Darkness Visible: A Memoir of Madness* (New York: Random House, 1990; repr. Vintage, 1992), 62.
2. Andrew Solomon, "Depression, Too, Is a Thing with Feathers," *Contemporary Psychoanalysis* 44, no. 4 (2008): 509–30.
3. Helen S. Mayberg, "Targeted Electrode-Based Modulation of Neural Circuits for Depression," *Journal of Clinical Investigation* 119, no. 4 (2009): 717–25.
4. Eric R. Kandel, "The New Science of Mind," *Gray Matter, Sunday Review, New York Times*, September 6, 2013.
5. Mayberg, "Targeted Electrode-Based Modulation."
6. Francisco López-Muñoz and Cecilio Alamo, "Monoaminergic Neurotransmission: The History of the Discovery of Antidepressants from 1950s until Today," *Current Pharmaceutical Design* 15, no. 14 (2009): 1563–86.
7. Ronald S. Duman and George K. Aghajanian, "Synaptic Dysfunction in Depression: Potential Therapeutic Targets," *Science* 338, no. 6103 (2012): 68–72.
8. Sigmund Freud and Josef Breuer, "Case of Anna O," in *Studies on Hysteria*, trans. and ed. James Strachey and Anna Freud (London: Hogarth Press, 1955).
9. Steven Roose, Arnold M. Cooper, and Peter Fonagy, "The Scientific Basis of Psychotherapy," in *Psychiatry*, 3rd ed., eds. Allan Tasman et al. (Chichester, UK: John Wiley and Sons, 2008), 289–300.
10. Aaron T. Beck et al., *Cognitive Therapy of Depression* (New York: Guilford Press, 1979).
11. Ibid.
12. Kay Redfield Jamison, *An Unquiet Mind: A Memoir of Moods and Madness* (New York: Alfred A. Knopf, 1995), 89.
13. Solomon, "Depression, Too, Is a Thing with Feathers."
14. Mayberg, "Targeted Electrode-Based Modulation."
15. Sidney H. Kennedy et al., "Deep Brain Stimulation for Treatment-Resistant Depression: Follow-Up After 3 to 6 Years," *American Journal of Psychiatry* 168, no. 5 (2011): 502–10.
16. Jamison, *An Unquiet Mind*, 67.
17. Jane Collingwood, "Bipolar Disorder Genes Uncovered," *Psych Central*, May 17, 2016, https://psychcentral.com/lib/bipolar-disorder-genes-uncovered/.

4. THE ABILITY TO THINK AND TO MAKE AND CARRY OUT DECISIONS: SCHIZOPHRENIA

For a general discussion on schizophrenia, see Steven E. Hyman and Jonathan D. Cohen, "Disorders of Thought and Volition: Schizophrenia," in Kandel et al., *Principles of Neural Science*, 1389–1401.

1. Elyn R. Saks, *The Center Cannot Hold: My Journey through Madness* (New York: Hyperion, 2007), 1–2.
2. Irwin Feinberg, "Cortical Pruning and the Development of Schizophrenia," *Schizophrenia Bulletin* 16, no. 4 (1990): 567–68.
3. Jill R. Glausier and David A. Lewis, "Dendritic Spine Pathology in Schizophrenia," *Neuroscience* 251 (2013): 90–107.
4. Daniel H. Geschwind and Jonathan Flint, "Genetics and Genomics of Psychiatric Disease," *Science* 349, no. 6255 (2015): 1489–94.
5. David St. Clair et al., "Association within a Family of a Balanced Autosomal Translocation with Major Mental Illness," *Lancet* 336, no. 8706 (1990): 13–16.
6. Qiang Wang et al., "The Psychiatric Disease Risk Factors DISC1 and TNIK Interact to Regulate Synapse Composition and Function," *Molecular Psychiatry* 16, no. 10 (2011): 1006–23.
7. Aswin Sekar et al., "Schizophrenia Risk from Complex Variation of Complement Component 4," *Nature* 530, no. 7589 (2016): 177–83.
8. Ryan S. Dhindsa and David B. Goldstein, "Schizophrenia: From Genetics to Physiology at Last," *Nature* 530, no. 7589 (2016): 162–63.
9. Christoph Kellendonk et al., "Transient and Selective Overexpression of Dopamine D2 Receptors in the Striatum Causes Persistent Abnormalities in Prefrontal Cortex Functioning," *Neuron* 49, no. 4 (2006): 603–15.

5. MEMORY, THE STOREHOUSE OF THE SELF: DEMENTIA

1. Larry R. Squire and John T. Wixted, "The Cognitive Neuroscience of Human Memory Since H.M.," *Annual Review of Neuroscience* 34 (2011): 259–88.
2. Eric R. Kandel, "The Molecular Biology of Memory Storage: A Dialogue Between Genes and Synapses," *Science* 294, no. 5544 (2001): 1030–38.
3. D. O. Hebb, *The Organization of Behavior: A Neuropsychological Theory* (New York: John Wiley and Sons, 1949).
4. Bengt Gustafsson and Holger Wigström, "Physiological Mechanisms Underlying Long-Term Potentiation," *Trends in Neurosciences* 11, no. 4 (1988): 156–62.
5. Elias Pavlopoulos et al., "Molecular Mechanism for Age-Related Memory Loss: The Histone-Binding Protein RbAp48," *Science Translational Medicine* 5, no. 200 (2013): 200ra115.
6. Ibid.
7. Ibid.
8. Franck Oury et al., "Maternal and Offspring Pools of Osteocalcin Influence Brain Development and Functions," *Cell* 155, no. 1 (2013): 228–41.
9. Stylianos Kosmidis et al., "Administration of Osteocalcin in the DG/CA3 Hippocampal Region Enhances Cognitive Functions and Ameliorates Age-Related Memory Loss via a RbAp48/CREB/BDNF Pathway" (in preparation).

10. Ibid.
11. Rita Guerreiro and John Hardy, "Genetics of Alzheimer's Disease," *Neurotherapeutics* 11, no. 4 (2014): 732–37.
12. R. Sherrington et al., "Alzheimer's Disease Associated with Mutations in Presenilin 2 is Rare and Variably Penetrant," *Human Molecular Genetics* 5, no. 7 (1996): 985–88.
13. Thorlakur Jonsson et al., "A Mutation in *APP* Protects against Alzheimer's Disease and Age-Related Cognitive Decline," *Nature* 488, no. 7409 (2012): 96–99.
14. Bruce L. Miller, *Frontotemporal Dementia*, Contemporary Neurology Series (Oxford, U.K.: Oxford University Press, 2013).

6. OUR INNATE CREATIVITY: BRAIN DISORDERS AND ART

1. Ann Temkin, personal communication, 2016.
2. Howard Gardner, *Multiple Intelligences: New Horizons*, rev. ed. (New York: Basic Books, 2006).
3. Benjamin Baird et al., "Inspired by Distraction: Mind Wandering Facilitates Creative Incubation," *Psychological Science* 23, no. 10 (2012): 1117–22.
4. Ernst Kris, *Psychoanalytic Explorations in Art* (New York: International Universities Press, 1952).
5. Bruce L. Miller et al., "Emergence of Artistic Talent in Frontotemporal Dementia," *Neurology* 51, no. 4 (1998): 978–82.
6. John Kounios and Mark Beeman, "The Aha! Moment: The Cognitive Neuroscience of Insight," *Current Directions in Psychological Science* 18, no. 4 (2009): 210–16.
7. Charles J. Limb and Allen R. Braun, "Neural Substrates of Spontaneous Musical Performance: An fMRI Study of Jazz Improvisation," *PLOS One* 3, no. 2 (2008): e1679.
8. Philippe Pinel, "Medico-Philosophical Treatise on Mental Alienation or Mania (1801)," *Vertex* 19, no. 82 (2008): 397–400.
9. Benjamin Rush, *Medical Inquiries and Observations, upon the Diseases of the Mind* (Philadelphia: Kimber and Richardson, 1812).
10. Cesare Lombroso, *The Man of Genius* (London: W. Scott, 1891).
11. Rudolf Arnheim, "The Artistry of Psychotics," *American Scientist* 74, no. 1 (1986): 48–54.
12. Thomas Roeske and Ingrid von Beyme, *Surrealism and Madness* (Heidelberg, Germany: Sammlung Prinzhorn, 2009).
13. Hans Prinzhorn, *Artistry of the Mentally Ill: A Contribution to the Psychology and Psychopathology of Configuration*, 2nd German ed., trans. by Eric von Brockdorff (New York: Springer-Verlag, 1995).
14. Ibid., 266.
15. Ibid., 265.
16. Ibid., vi.
17. Ibid., 150.
18. Ibid., 181.
19. Ibid., 160.
20. Ibid., 168–69.

21. Birgit Teichmann, Universität Heidelberg, personal communication, May 12, 2009.
22. Danielle Knafo, "Revisiting Ernst Kris' Concept of Regression in the Service of the Ego in Art," *Psychoanalytic Review* 19, no. 1 (2002): 24–49.
23. Kay Redfield Jamison, *Touched with Fire: Manic-Depressive Illness and the Artistic Temperament* (New York: The Free Press, 1993).
24. Nancy C. Andreasen, "Secrets of the Creative Brain," *The Atlantic*, July/August 2014, www.theatlantic.com/magazine/archive/2014/07/secrets-of-the-creative-brain/372299/.
25. Jamison, *Touched with Fire*.
26. Ruth Richards et al., "Creativity in Manic-Depressives, Cyclothymes, Their Normal Relatives, and Control Subjects," *Journal of Abnormal Psychology* 97, no. 3 (1988): 281–88.
27. Catherine Best et al., "The Relationship Between Subthreshold Autistic Traits, Ambiguous Figure Perception and Divergent Thinking," *Journal of Autism and Developmental Disorders* 45, no. 12 (2015): 4064–73.
28. Oliver Sacks, *An Anthropologist on Mars: Seven Paradoxical Tales* (New York: Alfred A. Knopf, 1995), 203.
29. Ibid.
30. David T. Lykken, "The Genetics of Genius," in *Genius and Mind: Studies of Creativity and Temperament*, ed. Andrew Steptoe (Oxford, U.K.: Oxford University Press, 1998), 15–37.
31. Francesca Happé and Uta Frith, "The Beautiful Otherness of the Autistic Mind," *Philosophical Transactions of the Royal Society B: Biological Sciences* 364, no. 1522 (2009): 1346–50.
32. Darold A. Treffert, "The Savant Syndrome: An Extraordinary Condition. A Synopsis: Past, Present, Future," *Philosophical Transactions of the Royal Society B: Biological Sciences* 364, no. 1522 (2009): 1351–57.
33. Allan Snyder, "Explaining and Inducing Savant Skills: Privileged Access to Lower Level, Less-Processed Information," *Philosophical Transactions of the Royal Society B: Biological Sciences* 364, no. 1522 (2009): 1399–1405.
34. Pia Kontos, "The Painterly Hand: Rethinking Creativity, Selfhood, and Memory in Dementia," Workshop 4: Memory and/in Late-Life Creativity (London: King's College, 2012).
35. Bruce L. Miller et al., "Enhanced Artistic Creativity with Temporal Lobe Degeneration," *Lancet* 348, no. 9043 (1996): 1744–45.
36. Wil S. Hylton, "The Mysterious Metamorphosis of Chuck Close," *The New York Times Magazine*, July 13, 2016.
37. Ibid.
38. Ibid.
39. Rudolf Arnheim, "The Artistry of Psychotics," in *To the Rescue of Art: Twenty-Six Essays* (Berkeley: University of California Press, 1992), 144–54.
40. Andreasen, "Secrets of the Creative Brain."
41. Jamison, *Touched with Fire*, 88.
42. Andreason, "Secrets of the Creative Brain."
43. Ibid.
44. Robert A. Power et al., "Polygenic Risk Scores for Schizophrenia and Bipolar Disorder Predict Creativity," *Nature Neuroscience* 18, no. 7 (2015): 953–55.

45. Ian Sample, "New Study Claims to Find Genetic Link Between Creativity and Mental Illness," *The Guardian*, June 8, 2015, www.theguardian.com/science/2015/jun/08/new-study-claims-to-find-genetic-link-between-creativity-and-mental-illness.
46. Andreason, "Secrets of the Creative Brain."

7. MOVEMENT: PARKINSON'S AND HUNTINGTON'S DISEASES

1. Charles S. Sherrington, *The Integrative Action of the Nervous System* (New Haven, CT: Yale University Press, 1906).
2. James Parkinson, "An Essay on the Shaking Palsy. 1817," *Journal of Neuropsychiatry and Clinical Neurosciences* 14, no. 2 (2002): 223–36.
3. Arvid Carlsson, Margit Lindqvist, and Tor Magnusson, "3,4-Dihydroxyphenylalanine and 5-hydroxytryptophan as Reserpine Antagonists," *Nature* 180, no. 4596 (1957): 1200.
4. A. Carlsson, "Biochemical and Pharmacological Aspects of Parkinsonism," *Acta Neurologica Scandinavica, Supplementum* 51 (1972): 11–42.
5. A. Carlsson and B. Winblad, "Influence of Age and Time Interval between Death and Autopsy on Dopamine and 3-Methoxytyramine Levels in Human Basal Ganglia," *Journal of Neural Transmission* 38, nos. 3–4 (1976): 271–76.
6. H. Ehringer and O. Hornykiewicz, "Distribution of Noradrenaline and Dopamine (3-Hydroxytyramine) in the Human Brain and Their Behavior in Diseases of the Extrapyramidal System," *Parkinsonism and Related Disorders* 4, no. 2 (1998): 53–57.
7. George C. Cotzias, Melvin H. Van Woert, and Lewis M. Schiffer, "Aromatic Amino Acids and Modification of Parkinsonism," *New England Journal of Medicine* 276, no. 7 (1967): 374–79.
8. Hagai Bergman, Thomas Wichmann, and Mahlon R. DeLong, "Reversal of Experimental Parkinsonism by Lesions of the Subthalamic Nucleus," *Science*, n.s., 249 (1990): 1436–38.
9. Mahlon R. DeLong, "Primate Models of Movement Disorders of Basal Ganglia Origin," *Trends in Neurosciences* 13, no. 7 (1990): 281–85.
10. D. Housman and J. R. Gusella, "Application of Recombinant DNA Techniques to Neurogenetic Disorders," *Research Publications—Association for Research in Nervous and Mental Disorders* 60 (1983): 167–72.
11. The Huntington's Disease Collaborative Research Group, "A Novel Gene Containing a Trinucleotide Repeat That Is Expanded and Unstable on Huntington's Disease Chromosomes," *Cell* 72 (1993): 971–83.
12. Stanley B. Prusiner, "Novel Proteinaceous Infectious Particles Cause Scrapie," *Science* 216, no. 4542 (1982): 136–44.
13. Stanley B. Prusiner, *Madness and Memory: The Discovery of Prions—A New Biological Principle of Disease* (New Haven, CT: Yale University Press, 2014), x.
14. Mel B. Feany and Welcome W. Bender, "A Drosophila Model of Parkinson's Disease," *Nature* 404, no. 6776 (2000): 394–98.

8. THE INTERPLAY OF CONSCIOUS AND UNCONSCIOUS EMOTION: ANXIETY, POST-TRAUMATIC STRESS, AND FAULTY DECISION MAKING

1. William James, "What Is an Emotion?" *Mind* 9, no. 34 (April 1, 1884), 190.
2. Aristotle, Lesley Brown, ed., and David Ross, trans., *The Nicomachean Ethics* (Oxford: Oxford University Press, 2009).
3. Sandra Blakeslee, "Using Rats to Trace Anatomy of Fear, Biology of Emotion," *New York Times*, November 5, 1996.
4. Edna B. Foa and Carmen P. McLean, "The Efficacy of Exposure Therapy for Anxiety-Related Disorders and Its Underlying Mechanisms: The Case of OCD and PTSD," *Annual Review of Clinical Psychology* 12 (2016): 1–28.
5. Barbara O. Rothbaum et al., "Virtual Reality Exposure Therapy for Vietnam Veterans with Posttraumatic Stress Disorder," *Journal of Clinical Psychiatry* 62, no. 8 (2001): 617–22.
6. Mark Mayford, Steven A. Siegelbaum, and Eric R. Kandel, "Synapses and Memory Storage," *Cold Spring Harbor Perspectives in Biology* 4, no. 6 (2012): a005751.
7. Alain Brunet et al., "Effect of Post-Retrieval Propranolol on Psychophysiologic Responding during Subsequent Script-Driven Traumatic Imagery in Post-Traumatic Stress Disorder," *Journal of Psychiatric Research* 42, no. 6 (2008): 503–6.
8. William James, *The Principles of Psychology*, vol. 2 (New York: Henry Holt and Company, 1913), 389–90.
9. Antonio R. Damasio, *Descartes' Error: Emotion, Reason, and the Human Brain* (New York: G. P. Putnam's Sons, 1994), 34ff.
10. Ibid., 43.
11. Ibid., 44–45.
12. Joshua D. Greene et al., "An fMRI Investigation of Emotional Engagement in Moral Judgment," *Science* 293 (2001): 2105–8.
13. Kent A. Kiehl and Morris B. Hoffman, "The Criminal Psychopath: History, Neuroscience, Treatment, and Economics," *Jurimetrics* 51 (2011): 355–97.
14. Ibid. See also L. M. Cope et al., "Abnormal Brain Structure in Youth Who Commit Homicide," *NeuroImage Clinical* 4 (2014): 800–807, and interview with Kent Kiehl in Mike Bush, "Young Killers' Brains Are Different, Study Shows," *Albuquerque Journal*, June 9, 2014.

9. THE PLEASURE PRINCIPLE AND FREEDOM OF CHOICE: ADDICTIONS

1. James Olds and Peter Milner, "Positive Reinforcement Produced by Electrical Stimulation of Septal Area and Other Regions of Rat Brain," *Journal of Comparative and Physiological Psychology* 47, no. 6 (1954): 419–27.
2. Wolfram Schultz, "Neuronal Reward and Decision Signals: From Theories to Data," *Physiological Reviews* 95, no. 3 (2015): 853–951.
3. Nora D. Volkow et al., "Dopamine in Drug Abuse and Addiction: Results of Imaging Studies and Treatment Implications," *Archives of Neurology* 64, no. 11 (2007): 1575–79.
4. Lee N. Robins, "Vietnam Veterans' Rapid Recovery from Heroin Addiction: A Fluke or Normal Expectation?," *Addiction* 88, no. 8 (1993): 1041–54.

5. N. D. Volkow, Joanna S. Fowler, and Gene-Jack Wang, "The Addicted Human Brain: Insights from Imaging Studies," *Journal of Clinical Investigation* 111, no. 10 (2003): 1444–51.
6. N. D. Volkow, George F. Koob, and A. Thomas McLellan, "Neurobiologic Advances from the Brain Disease Model of Addiction," *New England Journal of Medicine* 374, no. 4 (2016): 363–71.
7. Eric J. Nestler, "On a Quest to Understand and Alter Abnormally Expressed Genes That Promote Addiction," *Brain and Behavior Research Foundation Quarterly* (September 2015): 10–11.
8. Eric R. Kandel, "The Molecular Biology of Memory: cAMP, PKA, CRE, CREB-1, CREB-2, and CPEB," *Molecular Brain* 5 (2012): 14.
9. Jocelyn Selim, "Molecular Psychiatrist Eric Nestler: It's a Hard Habit to Break," *Discover*, October 2001, http://discovermagazine.com/2001/oct/breakdialogue.
10. Nestler, "On a Quest to Understand and Alter Abnormally Expressed Genes," 10–11.
11. Eric J. Nestler, "Genes and Addiction," *Nature Genetics* 26, no. 3 (2000): 277–81.
12. Eric R. Kandel and Denise B. Kandel, "A Molecular Basis for Nicotine As a Gateway Drug," *New England Journal of Medicine* 371 (2014): 932–43.
13. Yan-You Huang et al., "Nicotine Primes the Effect of Cocaine on the Induction of LTP in the Amygdala," *Neuropharmacology* 74 (2013): 126–34.
14. Kyle S. Burger and Eric Stice, "Frequent Ice Cream Consumption Is Associated with Reduced Striatal Response to Receipt of an Ice Cream–Based Milkshake," *American Journal of Clinical Nutrition* 95, no. 4 (2012): 810–17.
15. Nicholas A. Christakis and James H. Fowler, "The Spread of Obesity in a Large Social Network over 32 Years," *New England Journal of Medicine* 357 (2007): 370–79.
16. Josh Katz, "Drug Deaths in America Are Rising Faster Than Ever," *The New York Times*, June 5, 2017.

10. SEXUAL DIFFERENTIATION OF THE BRAIN AND GENDER IDENTITY

1. Norman Spack, "How I Help Transgender Teens Become Who They Want to Be," TED, November 2013, www.ted.com/talks/norman_spack_how_i_help_transgender_teens_become_who_they_want_to_be; Abby Ellin, "Elective Surgery, Needed to Survive," *The New York Times*, August 9, 2017.
2. David J. Anderson, "Optogenetics, Sex, and Violence in the Brain: Implications for Psychiatry," *Biological Psychiatry* 71, no. 12 (2012): 1081–89; Joseph F. Bergan, Yoram Ben-Shaul, and Catherine Dulac, "Sex-Specific Processing of Social Cues in the Medial Amygdala," *eLife* 3 (2014): e02743.
3. Dick F. Swaab and Alicia Garcia-Falgueras, "Sexual Differentiation of the Human Brain in Relation to Gender Identity and Sexual Orientation," *Functional Neurology* 24, no. 1 (2009): 17–28.
4. Deborah Rudacille, *The Riddle of Gender: Science, Activism, and Transgender Rights* (New York: Pantheon, 2005), 21–22.
5. Ibid., 23.
6. Ibid., 24.

7. Ibid., 27.
8. Sam Maddox, "Barres Elected to National Academy of Sciences," *Research News*, Christopher and Dana Reeve Foundation, May 2, 2013, www.spinalcordinjury-paralysis.org/blogs/18/1601.
9. Rudacille, *Riddle of Gender*, 28–29.
10. Caitlyn Jenner, *The Secrets of My Life* (New York: Grand Central Publishing, 2017).
11. Diane Ehrensaft, "Gender Nonconforming Youth: Current Perspectives," *Adolescent Health, Medicine and Therapeutics* 8 (2017): 57–67.
12. Sara Reardon, "Largest Ever Study of Transgender Teenagers Set to Kick Off," *Nature* News, March 31, 2016, www.nature.com/news/largest-ever-study-of-transgender-teenagers-set-to-kick-off-1.19637.
13. Swaab and Garcia-Falgueras, "Sexual Differentiation of the Human Brain."

11. CONSCIOUSNESS: THE GREAT REMAINING MYSTERY OF THE BRAIN

1. Hyosang Lee et al., "Scalable Control of Mounting and Attack by Esr1+Neurons in the Ventromedial Hypothalamus," *Nature* 509 (2014): 627–32.
2. Bernard J. Baars, *A Cognitive Theory of Consciousness* (Cambridge, U.K.: Cambridge University Press, 1988).
3. Stanislas Dehaene, *Consciousness and the Brain: Deciphering How the Brain Codes Our Thoughts* (New York: Viking, 2014).
4. Ibid.
5. C. D. Salzman et al., "Microstimulation in Visual Area MT: Effects on Direction Discrimination Performance," *Journal of Neuroscience* 12, no. 6 (1992): 2331–55; C. D. Salzman and William T. Newsome, "Neural Mechanisms for Forming a Perceptual Decision," *Science* 264, no. 5156 (1994): 231–37.
6. N. K. Logothetis and Jeffrey D. Schall, "Neuronal Correlates of Subjective Visual Perception," *Science*, n.s., 245, no. 4919 (1989): 761–63.
7. N. K. Logothetis, "Vision: A Window into Consciousness," *Scientific American*, September 1, 2006, www.scientificamerican.com/article/vision-a-window-into-consciousness/.
8. Timothy D. Wilson, *Strangers to Ourselves: Discovering the Adaptive Unconscious* (Cambridge, MA: Harvard University Press, 2002).
9. Timothy D. Wilson and Jonathan W. Schooler, "Thinking Too Much: Introspection Can Reduce the Quality of Preferences and Decisions," *Journal of Personality and Social Psychology* 60, no. 2 (1991): 181–92.
10. Benjamin Libet et al., "Time of Conscious Intention to Act in Relation to Onset of Cerebral Activity (Readiness-Potential): The Unconscious Initiation of a Freely Voluntary Act," *Brain* 106 (1983): 623–42.
11. Amos Tversky and Daniel Kahneman, "The Framing of Decisions and the Psychology of Choice," *Science*, n.s., 211, no. 4481 (1981): 453–58.
12. Daniel Kahneman, *Thinking, Fast and Slow* (New York: Farrar, Straus and Giroux, 2011).
13. A. D. (Bud) Craig, "How Do You Feel—Now? The Anterior Insula and Human Awareness," *Nature Reviews Neuroscience* 10 (2009): 59–70; Hugo D. Critchley

et al., "Neural Systems Supporting Interoceptive Awareness," *Nature Neuroscience* 7, no. 2 (2004): 189–95.

14. G. Elliott Wimmer and Daphna Shohamy, "Preference by Association: How Memory Mechanisms in the Hippocampus Bias Decisions," *Science* 338, no. 6104 (2012): 270–73.
15. Michael N. Shadlen and Roozbeh Kiani, "Consciousness As a Decision to Engage," in *Characterizing Consciousness: From Cognition to the Clinic?*, eds. Stanislas Dehaene and Yves Christen (Berlin and Heidelberg: Springer-Verlag, 2011), 27–46.

ACKNOWLEDGMENTS

I have benefited greatly from the wonderful critical insights of my publisher, Eric Chinski, who reshaped the book in a number of important ways. I am also grateful to my colleagues at Columbia: Tom Jessell, Scott Small, Daniel Salzman, Mickey Goldberg, and Eleanor Simpson, for their thoughtful and detailed reading of an earlier draft. I am again deeply indebted to my wonderful editor, Blair Burns Potter, who worked with me on three earlier books and once again brought her critical eye and her insightful editing to this book. Finally, I am much indebted to Sarah Mack for her editorial work and development of the art program, and to Pauline Henick, who patiently typed the many versions of this book and skillfully guided it to completion.

INDEX

Page numbers in *italics* refer to illustrations.

PERMISSIONS ACKNOWLEDGMENTS

Grateful acknowledgment is made for permission to reprint the following material:

Excerpts from *Madness and Memory*, copyright © 2014 by Stanley B. Prusiner, M.D. Reprinted by permission of Yale University Press.

Excerpts from *The Riddle of Gender: Science, Activism, and Transgender Rights*, copyright © 2005, 2006 by Deborah Rudacille. Used by permission of Pantheon Books, an imprint of the Knopf Doubleday Publishing Group, a division of Penguin Random House LLC. All rights reserved.

Excerpts from "The Best Way I Can Describe What It's Like to Have Autism," reprinted by permission of Eric McKinney.

Excerpts from *Depression, Too, Is a Thing with Feathers*, copyright © 2008 by Andrew Solomon; reprinted by permission of Routledge Publishing, a division of Taylor & Francis Group.

ILLUSTRATION CREDITS

Unless otherwise indicated, most illustrations were created or adapted by Sarah Mack or are in the public domain.

Brain illustrations were created by Terese Winslow.

p. 14: Photograph of Golgi stain. Picture by Bob Jacobs.

p. 34: Photograph of Uta Frith. Used with permission.

p. 37: Diagram of eye movement patterns. Used with permission from Springer; courtesy of Kevin A. Pelphrey.

p. 39: Diagram of theory of mind. Used with permission from Elsevier Books.

p. 47: Diagram of single-nucleotide variation and diagram of copy number variations. Courtesy of Chris Willcox.

p. 48: Photograph of child with Williams syndrome. Courtesy of Terry Monkaba.

p. 48: Photograph of child with autism. Courtesy of Ursa Hoogle.

p. 62: Photograph of Andrew Solomon. Used with permission from Andrew Solomon; courtesy of Timothy Greenfield-Sanders.

p. 78: Photograph of Kay Redfield Jamison. Used with permission.

p. 87: Photograph of Elyn Saks. Courtesy of USC Gould School of Law.

p. 109: Photograph of H.M.'s brain and an intact brain. Courtesy of Press et al.

p. 119: Photograph of an amyloid plaque and a neurofibrillary tangle in the brain. Picture by Nigel Cairns.

p. 120: Diagram of amyloid plaque creation. Courtesy of Chris Willcox.

p. 121: Diagram of tau protein misfolding. Courtesy of Chris Willcox.

p. 132: *Big Self-Portrait* © Chuck Close; courtesy of Pace Gallery.

p. 134: *Roy II* © Chuck Close; courtesy of Pace Gallery.

p. 134: Detail of *Roy II* © Chuck Close; courtesy of Pace Gallery.

p. 141: *The Flamingoes* by Henri Rousseau. Used with permission from Dennis Hallinan / Alamy Stock Photo.

p. 149: *Remorse, or Sphinx Embedded in the Sand* by Salvador Dalí. Used with permission from Gala-Salvador Dalí Foundation / Artists Rights Society (ARS).

p. 173: The brain of the fruit fly. Used with permission from Columbia University; courtesy of Pavan K. Auluck, H. Y. Edwin Chan, John Q. Trojanowski, Virginia M. Y. Lee, and Nancy M. Bonini.

p. 179: Diagram of the valence of emotion. Courtesy of Paul Ekman.

p. 185: Photograph of marine. Courtesy of the U.S. National Archives.

p. 191: Diagram reconstructing the iron bar's pathway through Gage's brain. Adapted, with permission, from H. Damasio et al. 1994.

p. 192: Diagram of the trolley problem. Courtesy of Luigi Corvaglia.

p. 206: Diagram of the brain's normal reward circuitry disrupted by addiction. Courtesy of Eric Nestler.

p. 220: Photograph of Ben Barres. Used with permission.